ARTIFICIAL INTELLIGENCE IN MANUFACTURING
Key to Integration?

GDI

Technology Assessment and Management 3

Previous volumes in series:

Volume 1:
ARTIFICIAL INTELLIGENCE:
Towards Practical Application
Th. Bernold and G. Albers (eds.)

Volume 2:
EXPERT SYSTEMS AND KNOWLEDGE ENGINEERING
Essential Elements of Advanced Information Technology
Th. Bernold (ed.)

ARTIFICIAL INTELLIGENCE IN MANUFACTURING
Key to Integration?

Proceedings of the Technology Assessment and Management Conference of the Gottlieb Duttweiler Institute Rüschlikon, Zürich, Switzerland, 7-8 November 1985

edited by

Thomas BERNOLD
Project Manager
Gottlieb Duttweiler Institute

1987

NORTH-HOLLAND · AMSTERDAM · NEW YORK · OXFORD · TOKYO

ISBN: 0 444 70154 0

Published by:

ELSEVIER SCIENCE PUBLISHERS B.V.
P.O. Box 1991
1000 BZ Amsterdam
The Netherlands

Sole distributors for the U.S.A. and Canada:

ELSEVIER SCIENCE PUBLISHING COMPANY, INC.
52 Vanderbilt Avenue
New York, N.Y. 10017
U.S.A.

PRINTED IN THE NETHERLANDS

INTRODUCTION

Thomas Bernold

Integration has become one of the main issues in manufacturing today. Experts believe that progress in this area will mark a decisive step into the future of production or with other words toward the factory of the future. Despite great efforts there is however still a considerable gap between the user requirements and the performance of today's systems.

Taking a comprehensive view of manufacturing from product design, process design, production, distribution to field service and reclamation, complexity of the processes and the multilayered interactions is an inherent quality. Experience, a set of "rules of thumb" or "standards of good practice" used to be the approaches to these problems which are often ill defined and under constrained. Conventional computing, however, usually fails.

Combinatorial complexity and non-linearity are the types of problems Artificial Intelligence (AI) attempts to solve. Kempf (1) defines AI in the context of manufacturing as a three stage problem solving technique. The first step is represented in the question:"What knowledge is available to overcome the combinatorial explosion?"; the second: "How can this knowledge be represented in the computer?"; and the third: "How can this knowledge be utilized by the computer?".

This selection of papers presented at the third international conference on Technology assessment and Management at the Gottlieb Duttweiler Institute in Rüschlikon/Zürich: ARTIFICIAL INTELLIGENCE IN MANUFACTURING gives an overview of the technology and trends in this specific area of application of AI. It opens with a survey of the techniques presently used or developed. A detailed description of the manufacturing processes then points out the user requirements and challenges for the further development of AI systems. A sample of 18 case studies from the areas of advanced robotics, expert systems, integration, process control design and simulation illustrate the wide spectrum of current applications. The proceedings close with the discussion of the impact on management organisation and training.

It is the aim of this book to bridge the gap between technical knowledge and the managerial and organizational questions regarding applicability and introduction of these systems. It seems reasonable to claim that a sound knowledge about the technical concepts and their potential will become increasingly important for all levels of management. Even small and medium size companies may feel an increasing pressure to adopt the new technologies since the large conglomerates will tend to integrate their subcontractors into an overall manufacturing concept.

PREFACE

ARTIFICIAL INTELLIGENCE IN MANUFACTURING
T. Bernold (editor)
Elsevier Science Publishers B.V. (North-Holland)

INDUSTRIAL APPLICATIONS OF ARTIFICIAL INTELLIGENCE

Mark S. Fox

Intelligent Systems Laboratory, Robotics Institute
Carnegie-Mellon University
Pittsburgh, Pennsylvania 15213
USA

1. INTRODUCTION

The purpose of this paper is to review current and future applications of Artificial Intelligence (AI) and Knowledge-Based systems to manufacturing. This is not a review of robotics technology, but focuses, instead, on manufacturing decision problems. Manufacturing, in this case, refers to the entire product life cycle:

- product design,
- production planning,
- production,
- distribution, and
- field service and reclamation.

The review focuses on where, at each point in the product life cycle, there are problems to be solved, where AI is currently being applied, and where it may be applied in the future.

2. WHAT IS ARTIFICIAL INTELLIGENCE

Artificial Intelligence is the science concerned with the creation of machine intelligence which is able to perform tasks heretofore only performed by people. Much of this machine intelligence is symbolic and heuristic. Artificial Intelligence deviated from the juggernaut of computer research in the early 50s by exploring how computers can be used for more than just numeric processing. Back during the days when languages like COBOL and FORTRAN were being defined, people at Carnegie-Mellon University and MIT were investigating the simulation of human problem-solving on a computer. Some of the first programs at that time were being applied to solving logic problems as found in the book "Principle Mathematics" by Whitehead and Russell (Newell & Simon, 1956). Problems such as chess, checkers, image understanding, etc. began to be investigated. As a matter of fact, one interesting set of problems that people chose to demonstrate the AI techniques could solve problems at the same level that humans could, were chosen from the intelligence test that we normally give to students to measure their IQ (Evans, 1968). It so happens that computers are very good at solving them. The difficulty in the development of machine intelligence lies in programming computers to perform common sense reasoning, i.e. reasoning about every-day occurrences which people find easy to do.

The types of problems which AI attempts to solve are non-linear and combinatorially complex (e.g., planning/scheduling, image understanding, etc.). The impact of their being non-linear is that there do not exist algorithms which will provide optimal solutions in polynomial time. Hence, the use of symbolic, heuristic knowledge, or "rules of thumb", play a major role in AI systems.

AI research can be divided into two basic categories. Knowledge representation is concerned with h o w to represent knowledge in a computer understandable form, so that systems can act in an intelligent manner. Consider a description of an activity that occurs on a factory floor.

> The milling operation precedes inspection. It is composed of two steps: setup and run. Setup takes an hour and piece time is 30 minutes. Two resources are required by the operation. A wrench and an operator. The wrench can be found in the tool crib and is only required during setup. The operation is performed in cost center 84.

The question is: How would one represent the knowledge embedded in that paragraph in a computer? Typically, a record in a data base which contains fields in some language is created. Describing the basics of an operation, there may be a field in a record which describes the tooling required, type of operator, who the operator is, setup time, run time, and the next operation in a sequence of operations to produce a product. The problem is: How does a computer understand the meaning of the record. The answer is that the program which uses it projects its own interpretation. The goal of knowledge representation research is to move away from ad hoc representations of knowledge to a semantics-based representation which identifies the levels of representation and their machine interpretation (Brachman, 1979). One level of representation, the conceptual level, provides a standard semantics which can be used across organizations and tasks, such as simulation, scheduling, accounting, etc. Figure 2-1 is an example of the representation of the knowledge in that paragraph. It is relational in form. Knowledge representation research focuses on the identification standard types of nodes and relations. Figure 2-1 divides the knowledge into two types: activity and state. A state description describes a snapshot of the world before an activity is performed. For example "cost center 84 possesses a wrench" is a state description. It must be true in order to enable the milling activity to occur. States and activities are linked via causal relations. A state describes what must be true of the world to enable an activity to occur. In addition, activities may be defined at multiple levels of abstraction. Milling is refined into two sub-activities: the setup and running of the machine. Lastly, we must represent time. Setup, in time, occurs before the milling run. Time is not absolute, it is relative. When describing the factory floor, it is atypical to use absolute time periods, instead, activities are described as preceding each other; and hence, once the time of one activity is determined the time of all the other activities related to it can be determined.

This is a small glimpse of what people in AI have been doing in the area of knowledge representation. For more details on conceptual level representations of activities see (Sathi, Fox & Greenberg, 1985).

The second category of AI is research. Problems are solved by performing search. Search in scheduling may be described as a search which incrementally builds alternative schedules. First, starting with a single order, it generates alternative first operations. Then, for each operation, it generates alternative machines on which to perform that operation, and for each machine it generates alternative queue positions, that is, times to perform that operation. There may be other alternatives such as alternative shifts and substitute tooling and fixtures, which expand the tree into a larger network. Once the first operation is fully defined, the search proceeds with the next operation that follows it. Search can be performed in a forward manner or in a backward manner, starting from the due date.

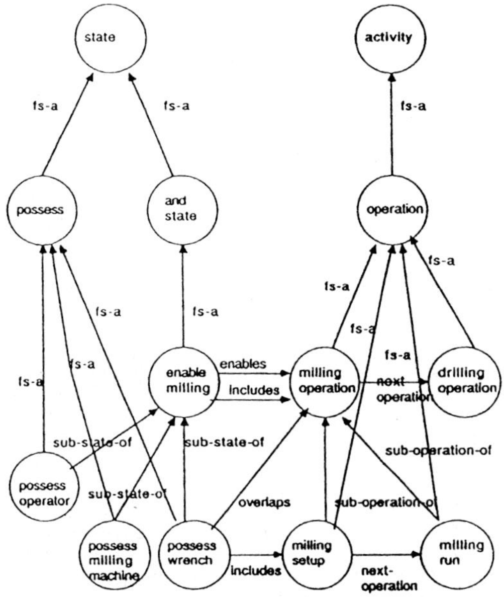

Figure 2-1: Activity Semantic Network

This is a very simple example of search. This type of search is similar to playing chess by considering all the alternatives. In a single factory which has 85 orders, 10 operations and only one substitutable machine, there are over 10^{800} alternative schedules that one could create. So, while this kind of search is an interesting, theoretical technique, practically, it is unusable because there are too many alternatives to consider. So, search has to be much smarter; in fact, search has to be structured in a manner that the search space is reduced from 10^{800} to something much smaller and manageable. The requirement to reduce the combinatorics of search has led to more complex and sophisticated search architectures. Figure 2-2 depicts the search architecture for the ISIS job shop scheduling system (Fox & Smith, 1984). ISIS has a hierarchical search architecture where it divides search at each level into three phases:

(1) pre-analysis of the search space to choose the operators which generate the search space,

(2) the search phase which performs the actual problem-solving, and

(3) post-analysis which analyzes the results of the search to identify whether it was successful or not and whether it should be performed again with a change in operators and constraints.

These phases are performed at each level within the search hierarchy. The first level chooses what order to schedule, the second level performs a search where it considers only a subset of the information in the factory, such as machine capacity, alternative operations, due dates and start dates, and tries to understand what the capacity is that is available, identify bottlenecks etc. The third level considers more information about the search space, substitutability of machines, tooling, shifts, constraints on how you perform your manufacturing, working in process, time constraints, how important it is to meet due dates, quality constraints, cost constraints, etc. The search results of higher levels constrain the processing of lower levels. Finally, the fourth level determines the time periods during which resources are allocated to operation. So, the system successively refines its schedules by starting from the aggregate and refining it via search to a detailed complete schedule by including more and more detail. Search-based systems like ISIS use sophisticated search techniques to reduce the size of the search space while still generating a good schedule.

The combination of representation and search are, in essence, the two legs of AI. It is upon these two legs that knowledge-based systems are constructed.

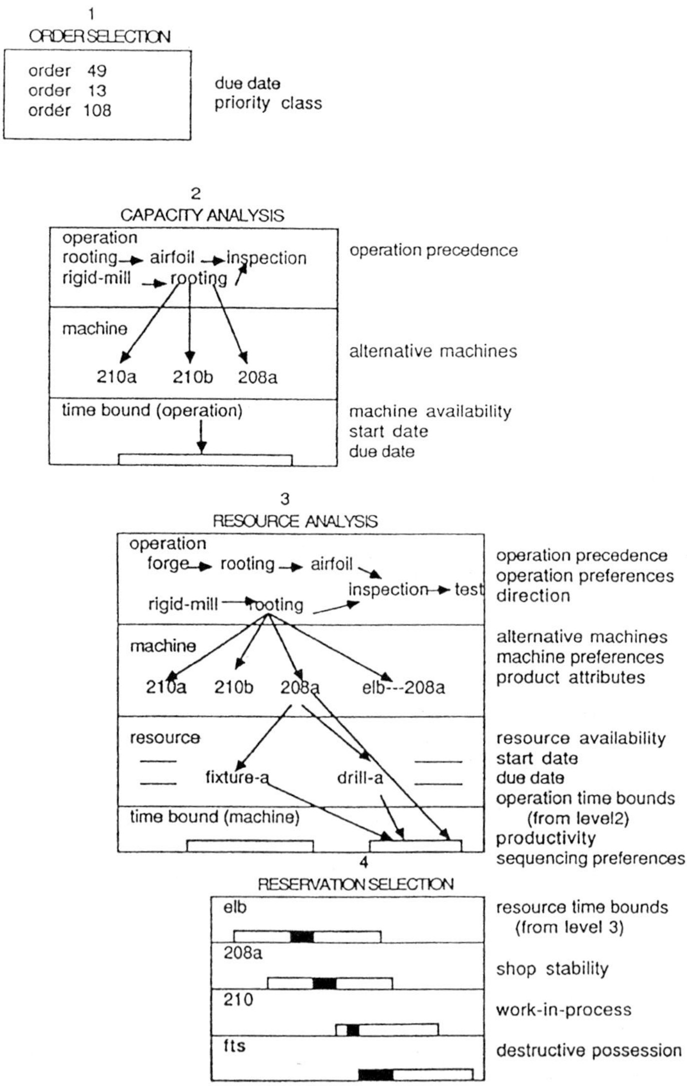

Figure 2-2: ISIS Search Hierarchy

3. ISSUES IN MANUFACTURING

There are a number of reasons why Artificial Intelligence may be important in the manufacturing environment. Consider the following issues. The scarcity of expertise is endemic to many corporations. Each time I visit a manufacturing organization there is at least one request of the following type: "Mark, we have an expert in a particular area and he/she is going to retire in two years. We want to save that knowledge for the organization." Not only do they want to capture that knowledge before the person retires, but they also want to distribute it throughout the organization. The issue of capturing scarce expertise and distributing it is a major problem faced by manufacturing organizations today. It occurs in tasks such as process planning, diagnosis of machinery, scheduling, and engineering design.

Another issue being faced in manufacturing is decision complexity. Decision complexity arises when there is a large number of choices from which to choose. They may be engineering choices of how to design a product, or they may be scheduling choices of how to produce the product. It is the case that flexible manufacturing systems are exacerbating the problem. The added flexibility on the factory floor provides the scheduling person with more alternatives, hence more choices of how to produce product. Today, making scheduling decisions in very rigid manufacturing environments is too difficult. As these environments become more flexible, the complexity of decisions increases. The same is true in engineering: the product complexity is increasing and, hence, design complexity increases.

Information is also becoming more complex. Many people use the term 'paperless factory' or 'paperless office'. The problem of how to get information from the factory floor on-line is now being solved. This creates another problem: how to reduce on-line information to only what is necessary for an individual to make a decision? Putting information on-line does not make it accessible to the individual decision maker. Reducing 1000 pages to a single screen full of pertinent information is an unsolved problem today. The intelligent reduction of information on an individual-by-individual basis is a long way off.

Decision timeliness is another issue which is coupled with decision complexity. Not only is decision complexity increasing, but the time to make a decision is decreasing. For example, programmable automation reduces the setup time of machinery, hence the scheduling decisions have to consider more alternatives in less time: how can faster, smarter decision-making systems be built?

Lastly, there is an issue of coordination. It is now known that design is intimately connected with production, distribution and field service. If a design is not optimized for manufacturability, assembly, distribution, or field service, then it will increase its cost of manufacturing with the possibility of reduced quality. The question is: how can designs be coordinated with all the down-stream activities?

These are some of the issues that affect the ability to increase the quality and the productivity of manufacturing operations. The question arises again as to whether Artificial Intelligence is a useful technology.

4. AI IN MANUFACTURING SURVEY

The results of a recent survey show the extent to which AI is being applied to manufacturing problems (Fox, 1986). The survey shows that in the 60s and 70s there was very little work being done (figure 4-1). But in the 1980s there are at least

68 systems in research, 38 in development, 9 in field test and 14 in production use. With a response of about 125 systems this represents about a quarter of the number of real systems being investigated today. There are closer to 500 systems that are under construction around the world today using Artificial Intelligence techniques.

The survey demonstrates that a number of people believe that AI will have an impact. The question is: where? In the following, I review the areas of impact by examining each phase of manufacturing.

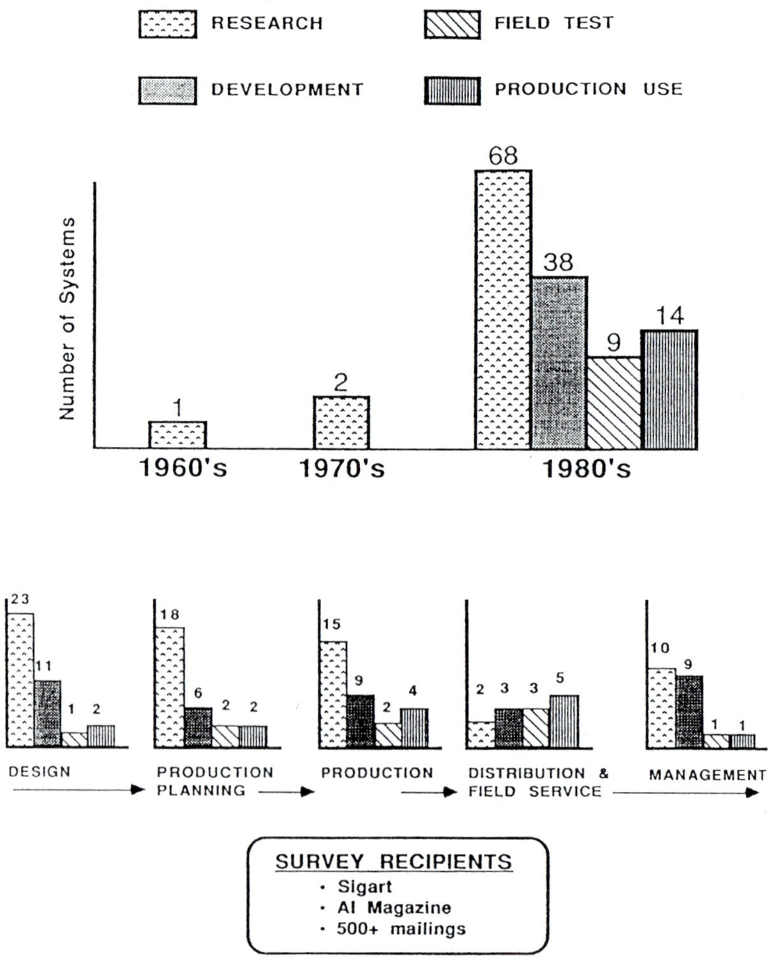

Figure 4-1: AI in Manufacturing Survey

5. AI IN DESIGN

The first phase of the product life cycle is design (figure 5-1). Design is composed of a product specification followed by the actual design and then its validation. In parallel, a number of management activities are performed. AI is being applied to each of these phases. Today, specifications are natural language descriptions written, in the case of commercial products, on a few pages, or in the case of military products, on thousands of pages. It is necessary to analyze these specifications to determine their completeness and consistency.

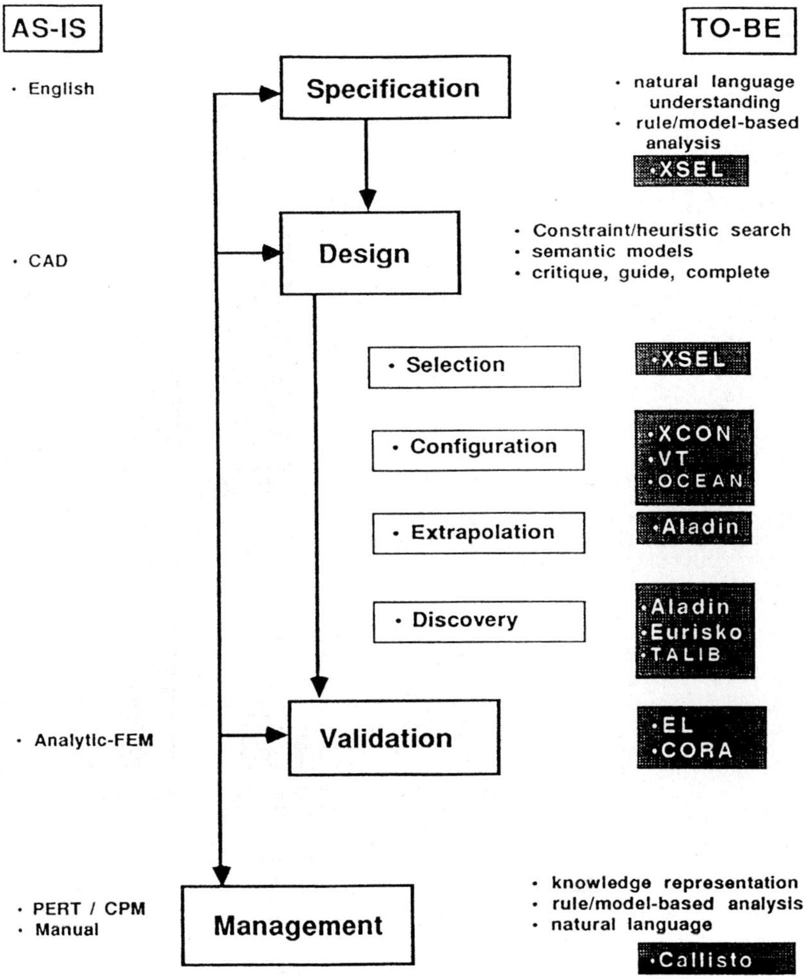

Figure 5-1: AI in Design

Artificial Intelligence has been applied to the acquisition and analysis of specifications. XSEL (McDermott, 1982) is an example of a system which works with the computer sales person to acquire customer product requirements. The XSEL system has a natural language interface for the specification of customer goals; it analyzes customer needs and queries the customer in order to complete the model of the customer's needs. This system is in production use. XSEL deals with a portion of the specification problem. The general ability to analyze complete specifications, identifying consistencies and incompleteness is beyond the state of the art but is something to which AI techniques can be applied.

Once the specification is received and its completeness and consistency verified, design may proceed. Design is one of the most difficult and creative of the manufacturing tasks. It requires not only education but much experience. Consequently it is a knowledge-intensive task which utilizes both analytic and heuristic knowledge, and therefore makes AI techniques appropriate.

There are at least categories of design. The first is selection, which maps functional requirements onto the attributes of an existing product line, hence selecting a product from an existing set. For example, XSEL maps user functional requirements onto product attributes and selects the appropriate products.

Configuration is a second type of design and is a little more difficult. The problem is to combine semi-finished components to form a final product. There are a number of successful configuration systems. XCON/R1 (McDermott, 1982) configures the VAX and PDP/11 lines of computers. It is in production use today. VT (Marcus et al., 1986) is another system which configures elevators and is in field test. OCEAN (NCR, 1985) is another computer configuration system. The XCON system has been in use for three years. It has configured over 100'000 computer orders at Digital Equipment Corp. and is, on the average, over 99 % correct, which exceeds the ability of human configurers. It is also faster, configuring systems in 3-5 minutes versus one hour to three hours for human configurers.

Extrapolation is the third type of design. Here, an existing product is altered to meet the customer's specifications. ALADIN (Farinacci et al., 1986) is an example of an extrapolation system. It takes the specification of the properties for an aluminum alloy, and alters an existing alloy's composition and thermal-mechanical processing, so that a new alloy which meets the customer's needs is defined.

Discovery is the most difficult type of design. It is similar to configuration in that it takes components and combines them. The difference is that configuration uses more semi-finished components, i.e., memory boards, CPU, busses, etc. for computers, whereas discovery begins with transistors and resistors. The distance in the functionality of the components from the end product determines whether it is discovery or configuration. There are systems which demonstrate the effectiveness of AI techniques on discovery problems. ALADIN can start just with aluminum to design an aluminum alloy. EURISCO (Lenat, 1983) is a discovery system which has been applied to the design of VLSI circuitry. TALIB (Kim & McDermott, 1983) is a system which designs electronic circuitry. All of these systems have been able to design/discover interesting functional systems out of basic primitives.

Once the design is complete, validation is used to verify that the form of the design behaves as defined by the functional specification. Simulation is used most often; for example, finite element analysis is used to test stresses in mechanical parts, and in electronic parts simulation is used to verify circuit behavior. The use of AI for validation differs in that it attempts to replicate what humans do when they validate a design: people have the ability to look at the structure of a design and infer functionality with minimal use of simulation. Two systems have been investigated in this area: the CONSTRAINTS system (Steele, 1980) infers functionality from the form of an electronic circuitry, and the CORA system (Westinghouse, 1985) verifies relay protection systems which protect power lines.

Lastly, AI is also being applied to the management of engineering projects. In particular, the CALLISTO system (Sathi et al., 1986) focuses on the management of product definition, and the activities performed to design the product. It provides a knowledge representation for the representation of dynamic objects and activities, and captures expertise for the management of activities.

6. AI IN PRODUCTION PLANNING

Production planning (figure 6-1) is the second stage of the manufacturing product life cycle. It takes a product design and definition of the production facilities as input. It forecasts customer demand, plans the process, lays out the facilities, specifies maintenance and trains the workers.

AI is being applied to each of the tasks. In the case of forecasting there are researchers in marketing groups who are looking at rule-based techniques for simulating market behavior. These systems go beyond the typical or techniques which tend to predict the future based on the past by having a greater understanding of the individual's needs. The ROME system (Kosy & Dhar, 1983) focuses on another part of the forecasting problem, the analysis of resource plans. It is an intelligent VISICALC^{tm} where the input is natural language and the system automatically analyzes the data to see if production goals are being met. If not, it figures out why.

A number of process planning systems are being investigated; ESP (Lindsay, 1985) has been developed for planning sheet metal production. XPSE (Cami, 1984), a continuation of the GARI system (Descotte & Latombe, 1981), is looking at the process planning of three-dimensional mechanical parts. Both are rule-based systems.

Process programming, for electronic applications programming, is straightforward and can be automated using conventional techniques. But for mechanical processes, such as the assembly of a carburetor, we do not know how to automate the programming of the assembly process. The application of AI to this problem is being investigated for the assembly of a power supply (Chang & Wee, 1985).

Another aspect of the process programming task is the selection of supporting resources such as cutting fluids for machining operations. GREASE (Mogush et al., 1986) is a system which selects cutting fluids for machining operations. It is now in field test.

Facility layout is concerned with how to organize production facilities based on process plans and product forecasts. The FADES system (Fisher, 1984) has been used to select the parameters of an Operations Research technique, which then determines the facility layout.

In maintenance design, AI is being used to develop maintenance procedure. RACE (Honeywell, 1985) is one such system now in field test.

Lastly, training is an important aspect of manufacturing. The impact of poorly trained workers can be devastating on production. Typically, training is performed manually, though simulations have been used for aircraft and nuclear power stations. What AI systems bring to the table is the ability to capture human expertise and using it to train new workers. It is also the case that expert simulations using Artificial Intelligence are being used. For example, Simulation Craft^{tm} (CGI, 1986) is an expert manufacturing simulator. It captures and provides scarce simulation, statistics and manufacturing expertise in support of the simulation life cycle: model specification, experiment design, and result analysis.

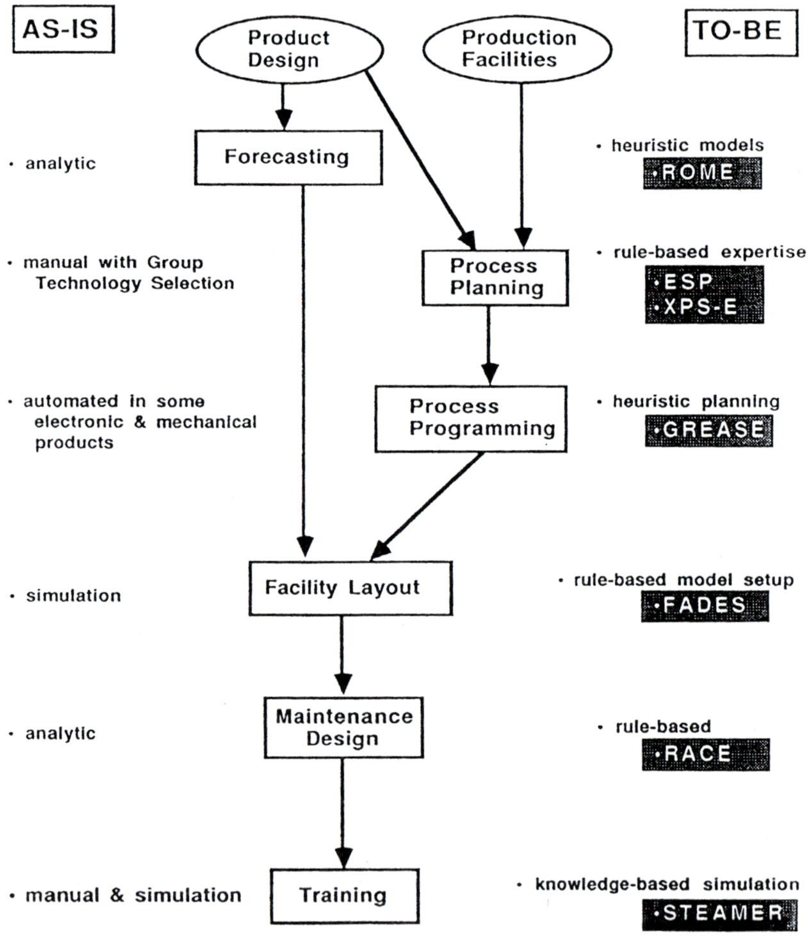

Figure 6-1: AI in Production Planning

7. AI IN PRODUCTION

Production (figure 7-1) is concerned with the planning, scheduling, management of the shop floor, control of cells, inspection of product and maintenance of processes.

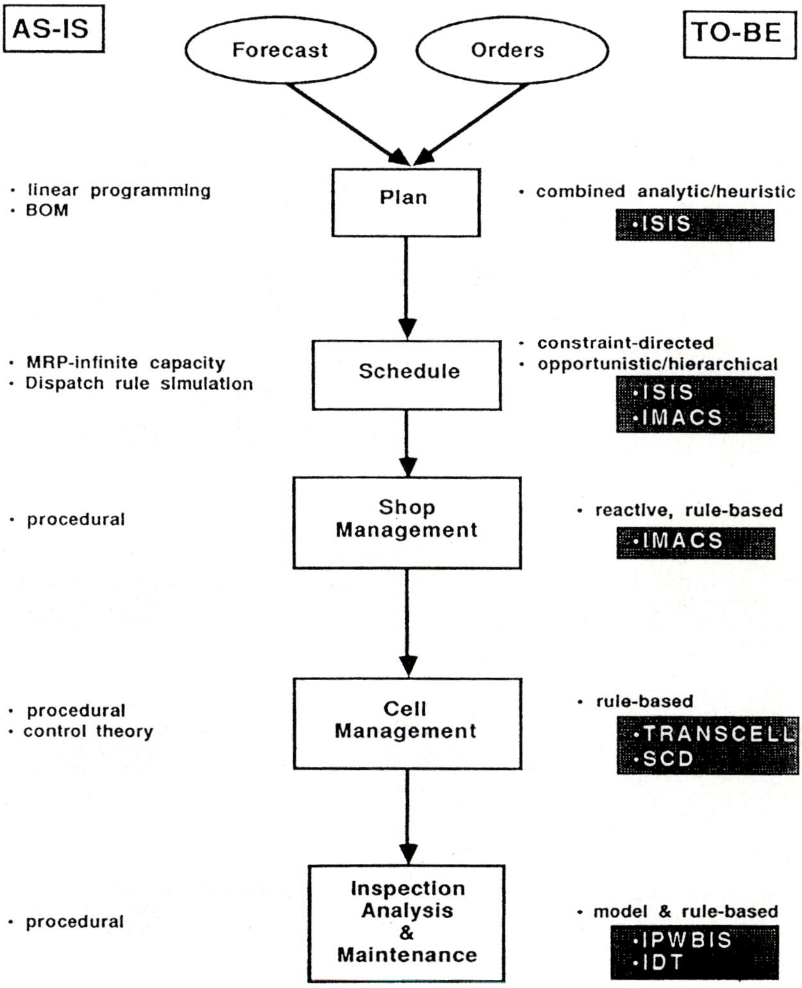

Figure 7-1: AI in Production

Scheduling is a combinatorial problem whose complexity, in many cases, exceeds the ability of human schedules. AI has been applied to scheduling. ISIS (Fox & Smith, 1984) is a system which uses constraint-directed search techniques for the scheduling of job-shops. Constraint-directed search uses constraints to reduce the combinatorics of the search resulting in schedules which satisfy some constraints while relaxing others.

IMACS (Haley et al., 1983) focuses on flow shop scheduling. The important concept in IMACS is its approach to managing a schedule. IMACS believes in Murphy's Law. Murphy's Law states that whatever can go wrong will go wrong. Rather than design the shop floor control systems to expect the schedule will be followed, it expects that deviations will occur and focuses on their detection and repair.

The same concept of looking for deviations is also used at the cell level by the TRANSCELL system (Bourne & Fox, 1984) and the SCD system (Tashiro et al., 1985) developed at Hitachi. These are rule-based systems which react to change and identify the next activity to perform.

For inspection, both rule-based and model-based techniques are used to diagnose computers and printed wire boards. IDT (Shubin & Ulrich, 1982) is a system for testing computers that are assembled, and IPWBIS (Seto et al., 1982) is a system for inspecting inner layers of printed wire boards and identifying where in the production process the error was introduced.

8. AI IN DISTRIBUTION AND FIELD SERVICE

Distribution and field service (figure 8-1) are the final phases of the manufacturing product life cycle. Distribution begins with the design of the organization: where to manufacture, how much to manufacture, how much to inventory, what transportation routes to use, what type of transportation to use, etc. Other distribution tasks are order entry, product installation planning, and diagnosis and repair of the product at the customer's site. AI again is being applied to each of these tasks. The INET system uses knowledge-based simulation techniques to model organization structures, simulate them and analyze automatically the results. XSEL is used for sales order entry. The output of XSEL is then provided to a distribution system, ILOG, which decides where to source the order, where to assemble, and how to ship it. Each of these systems uses rules to capture and apply expertise, and heuristic search to improve upon the solution.

Installation planning has also been attacked using AI. The XSITE system (McDermott & Steele, 1981) decides how to lay out the computer at the customer's site.

Lastly, diagnosis and repair is an area that has received a great deal of attention from AI. PDS (aka GENAID) (Fox et al., 1983) is a system which, using sensor input, performs diagnosis of steam turbines and generators, using sensor input. It has been in production use for more than a year. PDS is interesting because it has to worry about sensor inputs. The problem with sensor inputs is that the sensors degrade in many situations more quickly than the process itself. Diagnosis in this environment must determine whether the sensors are incorrect or whether the process itself is incorrect. That makes the problem a lot more difficult.

ACE (Stolfo & Vesonder, 1982) is a system which performs diagnosis of cable problems for telephone cabling. DOC is a system which performs diagnosis of computers and CATS is a system for the diagnosis of diesel locomotives. These are all rule-based systems which use heuristics, knowledge of situation and action. Current research in AI focuses on model-based reasoning, representing the physical structure and operating theory of the product, i.e. chemistry, physics, thermo dynamics, etc. For example, the DART system (Genesereth, 1982) diagnoses computers based upon their structure.

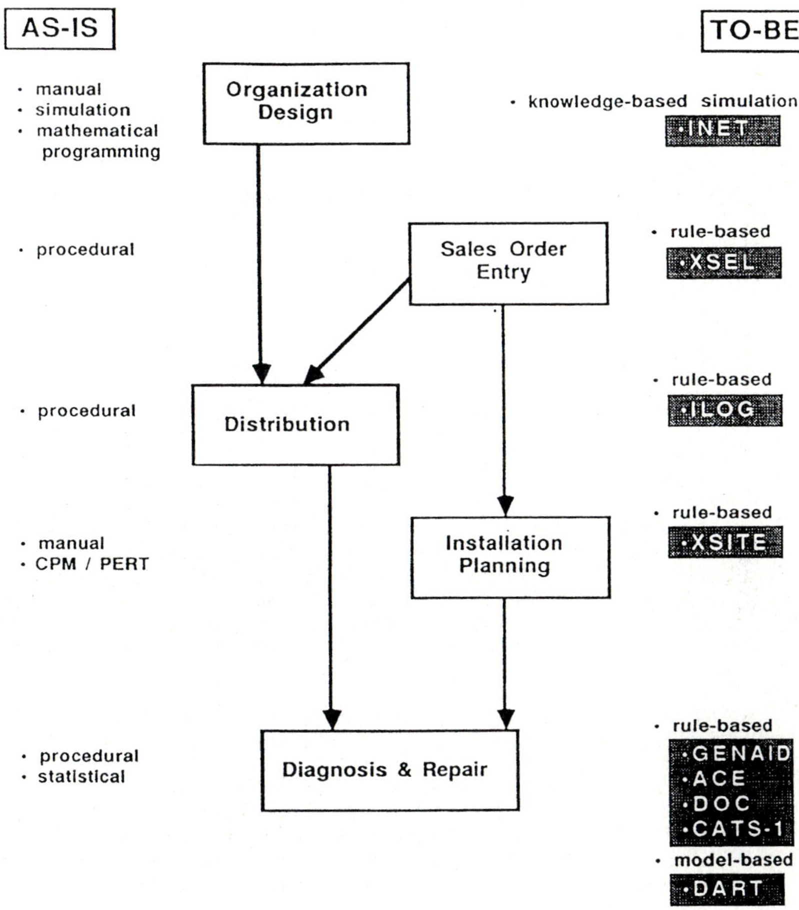

Figure 8-1: AI in Distribution and Field Service

9. FUTURE TRENDS

In the preceding, I have examined each phase of manufacturing: design, production planning, production, distribution, field service; and I have identified point applications of Artificial Intelligence. The question is: What is the future of applying Artificial Intelligence to manufacturing? What needs to be done is to take a system's view of the manufacturing process. One view is strategic: How can a product be designed so that it optimizes all the down-stream activities? I call this 'Design Fusion' (figure 9-1). The problem is to represent and utilize all the knowledge about planning, production, distribution and field service, during the design of the product, so that down-stream activities are optimized.

The second system's view is called 'Parts on Demand': How is an existing product pulled through the organization? Parts on demand decisions are concerned with tactical issues of: Where to get the product? How to get it more quickly? If the product does not exist, how can it be produced quickly? Or in the case where the design of the product no longer exists, how can it be reverse engineered? By taking a more global strategic view and tactical view of pushing new designs and pulling existing products within the organization, it becomes possible to see how spot applications of AI can be further optimized.

10. CONCLUSION

In conclusion, there exist a number of applications of AI in manufacturing today. They are beginning to impact manufacturing both on the shop floor and in engineering design. My expectations are that the number of systems will continue to increase at an even larger rate as corporations acquire more AI expertise and feel more comfortable about its application. More systems which capture scarce expertise and make it available throughout the organization will be created. Systems that enhance our problem solving by making better decisions more quickly will be created. Systems that integrate more knowledge about the factory floor and, hence, make better decisions will be created. There will be increased accessibility to these systems by people who are not computer-oriented through the use of natural language and explanation facilities.

The future appears to be rosy, but there are barriers looming along the road. The problem is that there are few people who really understand AI. There are people who have book knowledge of AI but do not understand AI from a point of view of actually building real, large systems. It so happens that AI is like a guild. It is a guild in a sense that there are masters, journeymen and apprentices. The problem is, there are not enough masters for the apprentices to work with. Because of this, many organizations will not have enough masters or any masters at all to guarantee a project's success. There will be a number of failures over the next five years, as corporations stub their toes in applying the technology. The possibilities of success may be enhanced if the corporation works with masters, starts small, lowers its expectations, and makes sure that the problem can be done by people before solving it with a computer. Only then will it be possible to successfully apply Artificial Intelligence to industrial problems. AI has a lot of promise but it must be applied cautiously.

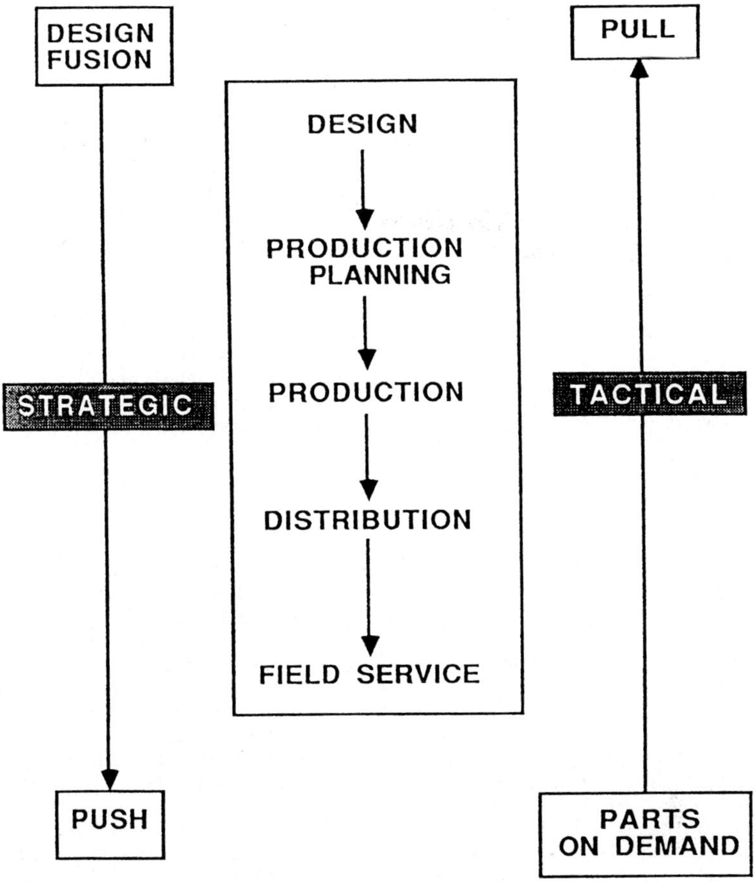

Figure 9-1: Future Trends

11. REFERENCES

Bourne, D., and Fox, M.S. (1984), "Autonomous Manufacturing: Automating the Job-Shop", Computer, Vol 17 No 9, pp. 76-88.

Brachman, R.J. (1979), "On the Epistemological Status of Semantic Networks", in: Associative Networks: Representation and Use of Knowledge by Computers, N.V. Findler (Ed.), pp. 3-50, New York, Academic Press.

Chang, K.H., and Wee, W.G. (1985), "A Knowledge Based Planning System for Mechanical Assembly Using Robots", Proceedings of the 22nd IEEE Design Automation Conference, pp. 330-336.

CGI (1986), "Simulation Craft", Carnegie Group Inc., Commerce Court at Station Square, Pittsburgh, PA 15219.

Descotte, Y., and Latombe, J.-C. (1981), "GARI: A Problem Solver That Plans How to Machine Mechanical Parts", Proceedings of the Seventh International Joint Conference on Artificial Intelligence, pp. 766-772, Vancouver, B.C.

Evans, T.G. (1968), "A Program for the Solution of Geometric-Analogy Intelligence Test Questions", in: Semantic Information Processing, M. Minsky (Ed.), Cambridge, MA, The MIT Press.

Farinacci, M.L., Fox, M.S., Hulthage, I., Rychener, M.D. (1986), "The Development of ALADIN, an Expert System for Aluminum Alloy Design", Artificial Intelligence in Manufacturing, Thomas Bernold (Ed.), Springer-Verlag, to appear.

Fisher, E.L. (1984), "Knowledge-Based Facilities Design", Ph.D. Thesis, School of Industrial Engineering, Purdue University, W. Lafayette, Indiana.

Fox, M.S., Lowenfeld, S., and Kleinosky, P. (1983), "Techniques for Sensor-Based Diagnosis", Proceedings of the International Joint Conference on Artificial Intelligence, August 1983, Los Altos, CA, William Kaufman Inc.

Fox, M.S., and Smith, S. (1984), "ISIS: A Knowledge-Based System for Factory Scheduling", International Journal of Expert Systems, Vol 1, No 1.

Genesereth, M.R. (1982), "Diagnosis Using Hierarchical Design Models", Proceedings of the National Conference on Artificial Intelligence, pp. 278-283, Los Altos, CA, William Kaufman Inc.

Haley, P., Kowalski, J., McDermott, J., and McWhorter, R. (1983), "PTRANS: A Rule-Based Management Assistant", Technical Report, Computer Science Dept., Carnegie-Mellon University.

Honeywell (1985), Personal communication with Honeywell Corporation.

Kim, J., and McDermott, J. (1983), "TALIB: An IC Layout Design Assistant", Proceedings of the National Conference on Artificial Intelligence, pp. 197-201, Los Altos, CA, William Kaufman Inc.

Kosy, D., and Dhar, V.S. (1983), "Knowledge-Based Support System for Long Range Planning", Technical Report, Robotics Institute, Carnegie-Mellon University, Pittsburgh, PA, December 1983.

Lenat, D.B. (1983), "Theory Formation by Heuristic Search. The Nature of Heuristics II: Background and Examples", Artificial Intelligence, Vol 21, No 1, pp.31-60.

Lindsay, K.J. (1985), "Frame-Based Knowledge Representation for Processing Planning", Proceedings of the First Annual Aerospace Applications of Artificial Intelligence, Dayton OH, Wright-Patterson Air Force Base Materials Lab, September 1985.

Marcus, S., Stout, J., and McDermott, J. (1986), "VT: An Expert Elevator Configurer", Technical Report, Computer Science Dept., Carnegie-Mellon University, Pittsburg, PA, to appear.

McDermott, J. (1982a), "R1: A Rule-Based Configurer of Computer Systems", Artificial Intelligence, Vol 19, No 1, pp. 39-88.

McDermott, J. (1982b), "XSEL: A Computer Sales Person's Assistant", in: Machine Intelligence 10, J.E. Hayes, D. Michie, and Y.H. Pao (Eds.), John Wiley & Sons.

McDermott, J., and Steele, B. (1981), "Extending a Knowledge-Based System to Deal with Ad Hoc Constraints", Proceedings of the Seventh International Joint Conference on Artificial Intelligence, Vancouver, B.C., August 1981.

Mogush, J.E., Carrega, D., Spirtes, P., and Fox, M.S. (1986), "Issues in Diagnosis and Treatment: A Case Study in Cutting Fluid Selection", Technical Report, Robotics Institute, Carnegie-Mellon University, Pittsburgh, PA, to appear.

Newell, A., and Simon, H.A. (1956), "The Logic Theory Machine: A Complex Information Processing System", IRE Transactions on Information Theory, Vol IT-2, No 3, pp. 61-79.

NCR (1985), Personal communication with National Cash Register Corporation.

Reddy, Y.V., Fox, M.S., Doyle, K., and Arnold, J. (1983), "INET: A Knowledge-Based Simulation Approach to Distribution Analysis", Proceedings of the IEEE Computer Society Trends and Applications, National Bureau of Standards, Washington, DC.

Sathi, A., Fox, M.S., and Greenberg, M. (1985), "Representation of Activity Knowledge for Project Management", IEEE Transactions on Pattern Analysis and Machine Intelligence, Vol PAMI-7, No 5, September 1985, pp. 531-552. Also appeared as Technical Report CMU-RI-TR-85-17, Robotics Institute, Carnegie-Mellon University.

Sathi, A., Morton, T., and Roth, S. (1986), "Callisto: An Intelligent Project Management System", AI Magazine, to appear.

Seto, J., Thibadeau, R., and Friedman, M. (1982), "Automatic Inspection for Printed Wiring", Technical Report CMU-RI-TR-82-16, Robotics Institute, Carnegie-Mellon University, Pittsburgh, PA.

Shubin, H., and Ulrich, J.W. (1982), "IDT: An Intelligent Diagnostic Tool", Proceedings of the National Conference on Artificial Intelligence, August 1983, Los Altos, CA, William Kaufman Inc.

Steele, G.L. (1980), "The Definition and Implementation of a Computer Programming Language Based on Constraints", (PhD Thesis), MIT Technical Report AI-TR-595, Cambridge, MA.

Stolfo, A., and Vesonder, G.T. (1982), "ACE: An Expert System Supporting Analysis and Management Decision Making", Technical Report, Computer Science Dept., Columbia University, New York, N.Y.

Tashiro, T., Komoda, N., Tsushima, I., and Matsumoto, K. (1985), "Rule-Based Control of Factory Automation Systems. An Approach with a Production System", Technical Report, Systems Development Lab., Hitachi Ltd.

Westinghouse (1985), Personal communication with Westinghouse Electric Corporation's Productivity and Quality Center.

TECHNOLOGY

ARTIFICIAL INTELLIGENCE IN MANUFACTURING
T. Bernold (editor)
Elsevier Science Publishers B.V. (North-Holland)

MANUFACTURING : A CHALLENGE FOR ARTIFICIAL INTELLIGENCE TECHNOLOGY

Jean-Paul Haton

CRIN-Nancy and INRIA

INTRODUCTION

Artificial Intelligence (AI) has experienced a rapid growth in industrial applications during the past few years. Significant realizations have been done in various domains : natural language processing, speech understanding, computer vision, expert systems, etc. They show the practical importance of AI even though present systems are still very limited in their capabilities.

The recent successes of AI are due to the conjunction of two factors. The first concerns the technological advances in the field of VLSI circuits that make it possible to design powerful systems at relatively low cost. The second, and most important factor, is the emergence of knowledge-based techniques. Knowledge, especially knowledge of experienced human experts, is a scarce and vital resource in industry. With these techniques it becomes possible to acquire expert knowledge and to use it for solving new classes of problems. An important feature of knowledge-based or expert systems, in comparison with classical software systems, is the clear distinction which is made between knowledge and the programs that manipulate it. This distinction makes it easier to incrementally build up large systems by progressively enriching the knowledge base corresponding to the problem. Another advantage is that such systems are to a certain extent able to explain their behaviour. This can be taken into account for the debugging of a system and for educating new experts.

AI will have a considerable impact in industry, particularly in the field of manufacturing. The general idea is to design more intelligent "robots", i.e. to implement more efficient and flexible connection between perception and action in the framework of computer aided manufacturing (CAM). We will try in this chapter to give an idea about the different ways AI technology will influence manufacturing. We will see that this influence corresponds to two broad categories which are highly interrelated :

- the communication between an automatic system and a human operator or the physical world. Natural language and speech are important media in the first case. We will briefly review the role of speech inputs and outputs in industrial applications. Computer vision is a fundamental mean for improving the interaction between a system and its environment. It is a basic component of factory automation,

- the intelligent control of systems. At this level expert systems will play an important role. We will present their specificities and discuss some open problems.

In fact manufacturing represents a formidable challenge for AI. The problems that arise are very difficult and several points remain unsolved in the fields of approximate reasoning, temporal reasoning, real time operation, knowledge representation, etc. But its economic and strategic importance is obvious.

INDUSTRIAL SPEECH INPUT AND OUTPUT

Speech can be an economic and natural mean for improving and simplifying man-machine communication. Although the problems are very difficult in their generality and not yet completely solved several limited applications have been proposed during the past ten years or so, especially in the industrial field. These developments concerne as well the synthesis of speech (make a machine talk) as the recognition of speech (make a machine listen). We will now successively examine these two aspects.

Speech synthesis : oral outputs are already used in various industrial applications that can be roughly grouped into three categories :

- safety of operation of a process : in a centralized control station of a large industrial process a huge amount of information is displayed. Speech synthesis is an efficient way for enhancing important information, e.g. in case of incidents. This technique has already been used in a number of applications, for instance in the control of electric power plan (at EDF in France) or of a rolling mill in steel industry at Solmer,

- security of factory workers : speech has been used for improving security, especially in case of emergency (emission of alarm messages, access to dangerous areas, etc.),

- user instruction guides : oral messages make it possible to assist a worker for carrying out a specific task when his eyes are already busy or when he is away (for instance maintenance of a system on a remote site).

Most of these applications use simple analysis-synthesis systems in which the speech signal has been recorded and stored in a compressed manner (synthesis by words). Such techniques are sufficient for small vocabulary situations. More sophisticated methods must be used for text-to-speech synthesis (synthesis by diphones, by rules, etc.). Present research in the field manly concerns low bit-rate synthesis and high quality, natural synthesis.

Speech Recognition : Present state of the art in automatic speech recognition allows the recognition of isolated or connected word for small vocabularies (some tens up to one hundred words), with a limited number of speakers [1]. In these cases dynamic programming algorithms provide a powerful tool for optimal, non linear time warping of speech patterns. That makes it possible to compensate for speech rate and rhythm distorsions. VLSI technology allows the low cost implementation of such systems on a restricted number of circuits (for instance NEC proposes a 3-chip word recognizer) with a very simple interface to a host computer. These systems are already able to handle practical industrial applications for which oral input present several advantages especially for manufacturing : the user can be busy looking at something or handling materials and he can be moving around, the terminal is simple and inexpensive compared to classical computer terminal, the input can be coupled with an audio feedback, etc.

Actual applications correspond to two different areas :

- data acquisition in the fields of sorting (parcels, flight luggages, etc.), inspection, quality control, CAD. Practical implementations have been made by car manufacturers (e.g. Renault, Volkswagen, or Ford with its warehouse shipping system that comprises 8 workstations, etc.) and in other industrial fields (e.g. Continental Can, Reynolds Metals and Westinghouse which uses its Series 100 voice systems in various inspection tasks [2] :
  . hybrid inspection of micro-electronic packages via microscopes,
  . inventory control on a semiconductor fusion test line,
  . final inspection of circuit boards.

An interesting application by General Electric concerns the use of both voice synthesis and recognition for aircraft engine inspection. That provides a direct link between the inspector and the assembly data base of the engine,

- control of machines : several experiments have been carried out for voice programming of numerically controlled machine tools [3]. Tests conducted by several industrial users (e.g. Orion in the USA or Oerlikon in W. Germany) resulted in substantial decrease of programming costs.

The oral control of robots has also been investigated. Present speech recognition systems severely limit the interaction on a word by word basis. It can be expected that the further use of a more or less constrainedlanguage with continuous speech facility will allows the implementation of an actual oral dialog between a man and a robot in conjunction with other communication media.

IMAGE PROCESSING AND COMPUTER VISION

Introduction

The use of images is increasing dramatically in our present civilisation, and especially for industrial applications. As a matter of fact artificial vision was among the first widespread uses of AI on the factory floor and the number of installed systems is growing rapidly, covering a large variety of applications even though they still feature only rudimentary vision capabilities. Such systems involve both AI and pattern recognition techniques which primarily consists of describing and classifying "patterns" (e.g. an object in a scene) by means of characteristic measurements on these patterns. Computer vision in its generality is a very difficult problem which, like other related domains such as speech recognition, is highly undeterministic : vision systems have to cope with a large number of ambiguities by an intensive use of all available knowledge sources. The simplest step in computer vision consists of processing binary (black and white) images. Many today industrial systems use binary imaging that turns out to be sufficient for simple tasks of inspection (e.g. counting simple pieces) and of object detection (e.g. in robot guidance if the robot has only to know whether an obstacle is present or not). More sophisticated tasks involve gray scaling where each pixel is assigned a numerical value proportional to the light intensity. 2D gray scale images require much more processing power but make it possible to do complex production or inspection tasks. Finally 3D vision is presently extensively studied in laboratories. 3D vision systems are required for interpreting scenes in such applications as bin-picking (in the most complex cases), sophisticated assembly line work and mobile robot guidance. The depth dimension can be obtained by active sensors like laser range finders or sonar systems. Research is also in progress concerning binocular 3D vision.

We can see that the range of industrial applications of computer vision is very wide. There are in fact two major fields of application :

- automated inspection. This field represents about 90 % of today installed systems. It ranges from simple inventory controls up to complex examination of objects for quality control. Half of today human product inspection activity is likely to be replaced by automatic vision systems during the next decade,

- automated guidance of systems. This field will certainly become more and more important in the future. It also covers tasks of various complexity : "pick-and-place" operations, welding, assembly, mobile robot guidance.

Before describing some practical applications we will now go a little bit further into the methodology of image processing and vision systems.

Methodology of computer vision [4]

Computer Vision is basically a hierarchical process, ranging from data acquisition up to the semantic description and interpretation of a scene, with close interactions and overlapping between the various processing levels. Figure 1 illustrates the overall architecture of a vision system. Let us consider in more details the various processing levels.

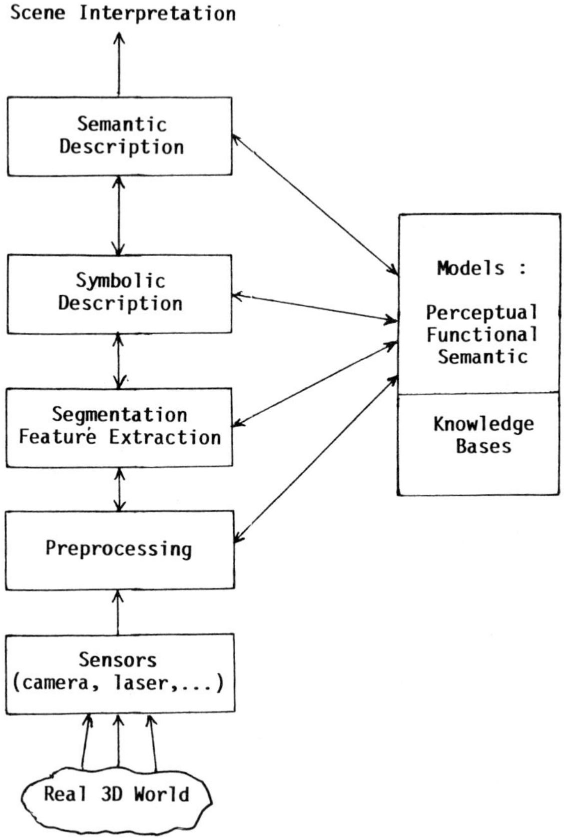

Figure 1
Architecture of a Computer Vision System

Preprocessing has been extensively studied in the framework of image processing. It is concerned with image restauration and enhancement in order to facilitate further processing. Techniques used at this level are essentially numerical and require an important processing power that can be obtained by massively parallel architectures.

The phase of attributes, or features, extraction is of primary importance since its role is to extract from the input data relevant information that will be processed by the other levels in order to solve specific problems. The idea is to build up algorithms capable of computing relations between the variations in intensity of pixels and the intrinsic properties (physical, geometrical, etc.) of the objects present in the scene. Several works, especially after Marr's ideas [5], have led to a fairly complete catalog of possible attributes :

- contours, made up of points featuring intensity discontinuities,
- textures which characterize local intensity variations within homogeneous zones delimited by contours,
- depth and distances between objects which are fundamental pieces of information for reasoning about the relative positions of objects,
- relative movements of objects in a scene.

The next step in the vision process consists of the symbolic manipulation of the scene representation. The goal is to obtain a symbolic description of a scene in terms of the relationships between surfaces or volumes which appear homogeneous with respect to one or several attributes. This description must be suitable for the further interpretation of the scene. It is usually incomplete and erroneous : this is a typical situation where AI may bring some solutions. Another problem, also directly related to AI, concerns the way in which the various knowledge pieces resulting from the symbolic manipulation will be represented in order to facilitate and speed up the interpretation.

The last level can be referred to as semantic manipulation level. It mainly consists of matching the symbolic description of a scene against stored models in order to interpret this scene according to the task to be done. Models represent for instance the structure of objects. If the scene has been properly segmented the matching process makes it possible to recognize objects even if they are distorted. Semantic nets have been widely used for representing such models. Other knowledge representation schemes proposed by AI have also proved useful, for instance frames. It is also often useful to control the matching process by some kind of rule-based reasoning.

The difficulty of computer vision makes it necessary to implement sophisticated AI search strategies in order to control combinatorial explosion of solutions. The paradigm of hypothesis-and-test, or prediction-verification, has often been used. It consists of emitting some hypothesis according to the context and the available knowledge and of verifying or cancelling it. In fact computer vision intimately combines algorithmic and heuristic AI techniques. AI is largely used in the so-called high level vision (corresponding to the symbolic and semantic manipulations) especially under the form of knowledge-based systems, but it can also bring some help for solving low level vision problems : for instance expert systems may be used for finding attributes on an image and/or segmenting this image. The more complex the scene to be interpreted and the task to be performed the more difficult and undeterministic the problem is. In the most difficult cases the key point is to take full advantage of data coming from different sensors (possibly not related to vision, e.g. contact sensors, etc.) and of all available knowledge sources. Specific architectures have to be designed for such a cooperation. The blackboard model, initially developed for speech understanding [6], provides a good framework for multi-sensor data fusion and multi knowledge sources problem solving. It has already been used for computer vision systems [7], [8] and it will certainly appear in complex systems for automated manufacturing.

<u>Examples of practical applications</u>

We have just seen that computer vision is very complex and still far from being completely solved. Vision systems are moreover expansive, even though technological advances in VLSI circuits will dramatically decrease the price of such systems. Therefore alternative solutions to vision are usually considered for industrial applications. For instance cheaper sensory mechanisms (e.g. ultrasonics) are preferred for rudimentary mechanical guidance. But vision becomes necessary for more sophisticated tasks and is therefore essential in factory automation.

Most of present industrial systems are concerned with low-end applications, essentially easy tasks in inspection and assembly.

As far as inspection is concerned [9], specialists usually agree that automated systems will increase productivity and improve quality. Practical applications can be found in very diverse industries : electronics (printed boards, photomasks of integrated circuits, chip inspection and alignment, inspection of electrical and electronic components), automobile (inspection of brake cylinders, camshaft, automobile frames, etc.), glass and ceramic, food, paper, pharmacy, metal processing (defect detection by surface inspection), etc.

Complex inspection and quality control tasks involve sophisticated human expertise, including visual expertise. Expert systems are therefore interesting to use at this level. Recent experiments concern the aluminium industry and the glass industry [10]. Combining vision with an expert system in these domains may lead to very efficient systems.

Companies are now targeting more ambitious tasks for the near future, in relation with the forthcoming new generations of computer vision systems. Typical examples concern "intelligent" welding robots able of following an edge between two parts or semi-automatic chip placement systems for pc boards assembly, a first step toward fully automated assembly lines.

Future Trends

The design of advanced computer vision systems is highly dependent of ongoing research efforts in various domains, especially in conjunction with AI. Crucial points concern first architectural aspects : efficient and cost-effective systems necessitate custom integrated circuits implementing the algorithms of low-level vision and also new architectures for the symbolic manipulations of high-level vision. At a conceptual level the automatization of vision represents one of the most challenging tasks of AI, particularly in the representation and use of knowledge. There is still a considerable effort to be done in order to design systems that could compare with human vision capabilities but present and forthcoming systems, although limited, will already have an important impact in industry.

PLANNING AND CONTROL

General Features

The control of a technical process is a fundamental task in manufacturing that is usually performed by a human operator assisted by automatic systems based on control theory algorithms. Several trends can be presently identified in the field :

- the task of the operator is becoming exceedingly complicated since the systems are becoming larger and larger and the number of variables and data to control too numerous,

- the complexity of the systems, and then of the operator task, is also rapidly increasing,

- there is a seamingly common tendancy to try to design fully automatic, unmanned plants or, at least, sections of a plant.

These trends raise very difficult problems that are almost impossible to solve by classical techniques. A key factor for successfully handling these problems is to use knowledge about the process, as human experts do. The technology of knowledge-based systems can therefore help getting some elements of solution in automated process control.

Almost all is still to be done and in fact this domain represents an important challenge for AI since the difficulties are far bigger than the ones encountered in present expert systems, e.g. for diagnostic tasks. These difficulties concern the necessity of operating in a real time mode, the prominent importance of temporal reasoning (cf. next paragraph) and also the fact that the number of data to be processed is very large (typically from some hundreds up to several thousand).

Three levels of knowledge are generally used by a human expert in a task of process control, i.e. "skills" corresponding to a surface knowledge of reflex reactions, rules for standard reasoning cases and deep knowledge for difficult, uncommon cases. Skills and rules are used in most present systems. However it will be necessary to take into account deep knowledge reasoning in order to design expert systems able to go beyond present intelligent assistants to operators.

Another key issue which is common to most industrial problems is that an expert system is not just an independent software which is put besides an existing process (e.g. a production line, a factory, a power plan, etc.). Rather the expert system must be intimately tight to the various elements of the process. Figure 2 shows the basic components of a knowledge based systems (i.e. the interpreter or inference engine, the knowledge base, the knowledge acquisition module and the explanation module) and their interaction with the outside world :

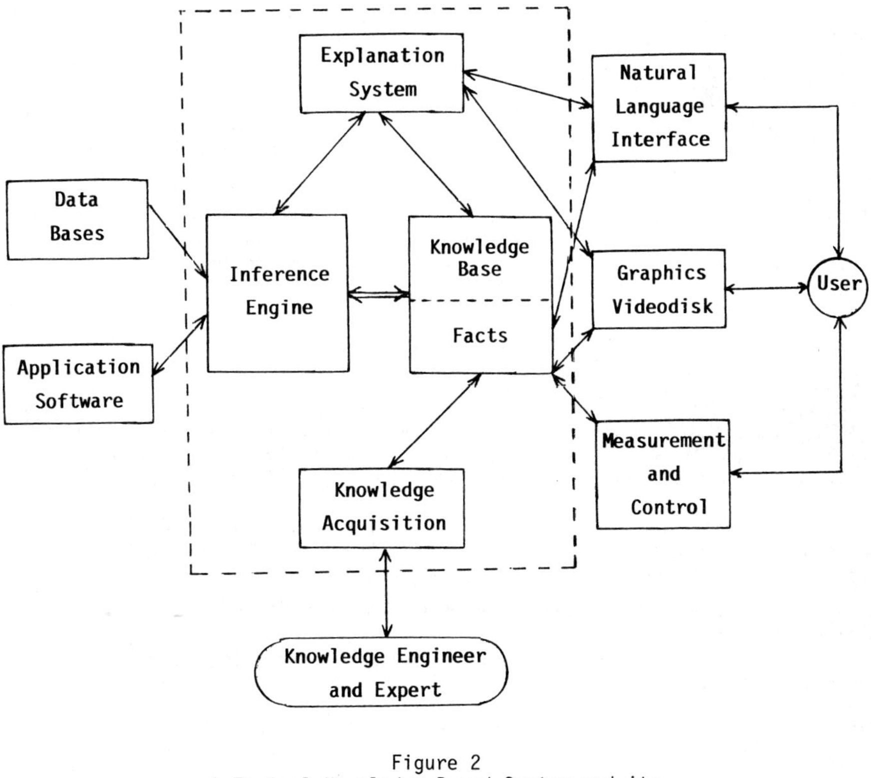

Figure 2
A Typical Knowledge-Based System and its Operational Environment

- with the user (that can be either a human operator or the process itself in case of closed loop operation). User interfaces include natural language processing capabilities as well as advanced visual displays (graphics, images) and, in case of closed loop functioning, procedures for automatic data acquisition and processing and feedback control,

- with existing data bases storing information about the process to be controlled, its history and all relevant information,

- with external applications software packages that can be called on by the interpreter for carrying out specific computations necessitated by the reasoning process.

Specific tasks that have to be considered in process control range from signal interpretation, diagnosis and repair to control. These tasks can be divided into four main categories of activity, i.e. :

- monitoring of the production process,

- diagnosis and repair (that includes the analysis of a fault). AI can be very useful in this area where knowledge is available from experts and can be relatively easily applied,

- fault prevention, a task which is usually badly taken into account since it necessitates very skilled experts. It consists of predicting forthcoming undesirable events by examining the behaviour of a system over a period of time (temporal reasoning) and of proposing actions which might prevent such events.

- modification of the process according to available resources at a certain time. This is a difficult operation which is very seldom undertaken. Planning techniques used in AI (e.g. in robotics) are necessary at this point.

A related domain concerns the problem of job shop scheduling. AI techniques have been used at this level for capturing the various kinds of knowledge necessary for establishing the schedule of the different phases for a given task. Expert systems prove useful for establishing and modifying heuristic rules that must be taken into account [11].

A relatively simpler planning task is related to the determination of system or product configurations. Expert systems have been built for configuring computer systems (XCON at DEC [12], OCEAN at NCR), and the principles can be extended to other manufacturing domains (e.g. the VT system used by Westinghouse for configuring elevators).

In the general framework of a man-machine interface (we don't consider the unmanned process control situation which is almost impossible with today technology) expert systems can be integrated in two ways which are not always exclusive :

- the expert system works in an advisory mode by giving advices to the human operator who in turn interacts with the process,

- the expert system appears as an intelligent interface between the process and the operator.

Current researches in the field have shown the usefulness of these two modes in a wide variety of applications, ranging from simple assembly line up to large integrated systems. Although practical systems are only emerging in real situations we will now examine on typical examples the problems encountered in process control expert systems design.

Expert Systems in Process Control

It does not exist presently any expert system capable of controlling large industrial systems in every day situations but limited realizations have been however implemented for various application domains.

We have already mentioned some of the problems that arise when developping systems in this area. Real time operation can be achieved through the use of high speed processors and parallel architectures. This point is not always crucial since the notion of real time is relative to a given process. Industrial processes may have slow variations and in that case the response time of the expert system becomes less critical.

A much more important issue concerns the necessity of temporal reasoning. Expert systems that are to monitor or to make diagnosis about dynamic systems must be able to take into account time-varying situations. Some limited attempts have been done to solve this problem. For instance VM is an expert system that interprets clinical data about patients in order to monitor a mechanical breathing assistance system [13]. In this application the reasoning model must be able to take into account data about the patient over time and to take decisions according to the progression of the disease. Knowledge in VM is represented under the form of production rules and its reasoning is limited to events which are adjacent in time. This is a highly restricted way of handling temporal reasoning and it is obvious that an important work is to be done at that level. However VM's model constitutes a good basis for control of industrial processes since the two problems are similar even though the number of measurements to be processed at each time sample is far lower in VM (some tens).

Another problem in process control expert systems also concerns reasoning. The nature of the data to be processed and of the human expertise makes it necessary to implement some kind of approximate reasoning. Several experiments have been conducted on the use of fuzzy logic in industrial process control [14]. One of the first successfull application on a full scale industrial process concerned a closed loop fuzzy controller of a rotary cement kiln in Denmark [15]. Fuzzy reasoning is not the only way for dealing with uncertainty but it provides a mean for expressing the linguistic terms used by a human skilled operator (for instance the expert will rate the numerical value of a measured variable with symbolic terms ranging from drastically low, low, slightly low, OK, etc. up to drastically high).

A limited number of expert systems development tools feature some of the above characteristics and are thus well suited for the process control field. One example is PICON, a real time expert system shell of LMI running on the Lambda Lisp Machine. It has been used for building intelligent alarm advisors for human operators. Its use in closed loop, without operator, is not envisaged before several years. Similar applications of expert systems that support human operators in diagnosis problems have also been extensively studied for nuclear power plants, a typical example of continuous process, for instance in the USA [16], in Japan [17] or in France [18].

The choice of a particular problem is usually not easy to do. Possible applications include :

- automatization of complete processes when all necessary knowledge is available and the technology exists,
- control of changes in the process state, e.g. start-up, change of operational mode (for instance a degraded mode in case of problem),
- alarm analysis,
- fault diagnosis and repair, a domain in which several systems have reached a high level of confidence as well for electronic circuits as for a locomotive (DELTA at General Electric [19]), a steel rolling mill or a steam engine (GENAID at Westinghouse).

Present Issues

Knowledge-Based techniques will certainly be increasingly involved in the management of industrial processes, even though present technology has not yet advanded to the point where complex situations can be handled.

Further research is needed, and is presently in progress, in some key directions. The following is a non-exhaustive list of such directions :

- the use of deep knowledge about the domain of expertise, in opposition with surface reasoning system, for instance in fault diagnosis [20]. That makes it possible to have a better understanding of the task and thus to make better diagnosis. Another interesting idea in the same field consists of reasoning not only on the data provided by sensors connected to the physical process but also on the sensors themselves in order to prevent errors in data collection [21],

- the representation of procedural knowledge in conjunction with declarative knowledge. This issue has been studied for several years. It is particularly useful in domains like process control where part of the available knowledge is under a procedural form. A related point consists of implementing a procedural control of a knowledge based system [22],

- the modeling of qualitative reasoning about physical processes [23]. This issue is concerned with the description and modeling of physical systems and of their changes, including commonsense knowledge about these systems.
An important motivation for applying AI techniques in manufacturing is to improve productivity. That requires better understanding of the underlying physical processes in order to carry out more sophisticated reasoning similar to the human expert (e.g. rely on different models of a process and choose the best one according to the actual situation). This point is directly related to the need of deep knowledge mentioned before,

- the design of specific architectures for complex process control. Such architectures will feature some distributuve aspects. A first reason is that real time requirements can be partly solved by distributing the reasoning over several concurrent processors ; that makes it also possible to examine multiple competing hypotheses at the same time. A second reason is related to the fact that data to be processed come from a (potentially) large number of sensors. Multiple data fusion can be envisaged through a distributed, cooperative association of processors. Finally the expert system must take into account several knowledge sources and make them cooperate (e.g. expertise of a human expert, models of the process, commonsense knowledge coming from qualitative reasoning, etc.). Several distributed architectures have been proposed and implemented in AI. The blackboard model that we have already encountered in computer vision and models derived from it appear as very attractive in this framework. Several experiments have proved their usefulness for signal interpretation [24] and for multi-sensor data fusion. They will certainly be more extensively developed for advances process control applications.

CONCLUSION

Artificial Intelligence is now entering the field of industrial applications and its impact in manufacturing will be of primary importance in the future. That constitutes an important challenge for AI since the problems which arise are very difficult. The limited realizations which are now operational already give an indication about forthcoming systems which will appear when adapted architectures are available and the knowledge-based tools more efficient.

We have shown that the impact of AI is twofold :

- on the communication of a system with the outside world (human operator and/or industrial process). Oral input-output and computer vision have been presented at that level,

- on the intelligent control of complex systems. Research and development on expert systems for process control have been presented.

AI is a mean for making robots intelligent. In the longer term robots will only be elements of computer integrated manufacturing systems including CAD/CAM, communications, controls, etc. AI will certainly be used in these various elements but it will at first be of primary importance in linking these elements together.

REFERENCES

[1] Haton, J.P., Speech Recognition and Understanding, Invited paper, 6th Int. Conf. on Pattern Recognition, Mϋnich,(1982) .

[2] Mandel, A.F., New System Overcomes Major Barriors to Voice Data Collection, Speech Technology, 3, n° 1, (1985) 72-75.

[3] Haton, J.P., A Practical Application of a Real Time Isolated Word Recognition System Using Syntactic Constraints, IEEE Trans. ASSP, 22, n° 6, (1974) 416-419.

[4] Ballard, D. and Brown, C., Computer Vision, (Prentice-Hall, 1982).

[5] Marr, D., Vision, Freeman, 1982.

[6] Lesser, V. et al., Organization of the HEARSAY II Speech Understanding System, IEEE Trans. ASSP, 23, n° 1, (1975) 11-23.

[7] Hanson, A.R. and Riseman, E.M., Visions : A Computer System for Interpreting Scenes, in : Hanson, A.R. and Riseman, E.M. (eds.), Computer Vision Systems (Academic Press, 1978).

[8] Nazif, A.M. and Levine, M.D., Low Level Image Segmentation : An Expert System, IEEE Trans. PAMI, 6, n° 5, (1984) 555-577.

[9] Chin, R.T., Automated Visual Inspection Techniques and Applications : A Bibliography, Pattern Recognition, 15, n° 4, (1982) 343-357.

[10] Mitra, S.K., Parker, J.M., An Expert System for Identifying Defects, Stones and Cord in Glass, Proc. ECAI-84, North-Holland,(1984).

[11] Le Pape C., Sauve B., SOJA : un système d'ordonnancement journalier d'atelier, 5èmes journées internationales Systèmes Experts et Applications, Avignon, (1985).

[12] Mc Dermott J., R1 : A Rule-Based Configurer of Computer Systems, Artifical Intelligence, 19, n° 1, (1982).

[13] Fagan, L.M. et al., Representation of Dynamic Clinical Knowledge : Measurement Interpretation in the Intensive Care Unit, Proc. 6th IJCAI, 1979.

[14] Mamdani, E.H. and Gaines, B.R., editors, Fuzzy Reasoning and its Applications, Academic Press, (1981).

[15] Martin Larsen, P., Industrial Applications of Fuzzy Logic Control, in |9|.

[16] Nelson, W.R., REACTOR : An Expert System for Diagnosis and Treatment of Nuclear Reactor Accidents, Proc. AAAI-2, Pittsburgh, (1982).

[17] Yamada, N. and Motoda, H., A Diagnosis Method of Dynamic System Using the Knowledge on System Description, Proc. IJCAI, (1983).

[18] Ancelin, J. et al., Couplage d'un système expert et d'une simulation pour l'étude du pilotage d'un processus, Actes du congrès Cognitiva, Paris, (1985).

[19] Kinnucan, P., Computers That Think Like Experts, High Technology, (January 1984), 30-42.

[20] Davis, R. et al., Diagnosis Based on Description of Structure and Function, Proc. AAAI-2, Pittsburgh, (1982).

[21] Fox, M, et al., Techniques for Sensor-Based Diagnosis, Proc. IJCAI, (1985).

[22] Gallanti, M, et al., Representing Procedural Knowledge in Expert Systems : an Application to Process Control, Proc. IJCAI, (1985).

[23] Bobrow, D.G., Qualitative Reasoning about Physical Systems : An Introduction, Artificial Intelligence, 24, (1984) 1-5.

[24] Nii, H.P. et al., Signal-to-Symbol Transformation : HASP/SIAP Case Study, AI Magazine, 3, (1982) 23-35.

APPLICATIONS

ARTIFICIAL INTELLIGENCE IN MANUFACTURING
T. Bernold (editor)
Elsevier Science Publishers B.V. (North-Holland)

GENERAL DESIGN THEORY AND ARTIFICIAL INTELLIGENCE

Hiroyuki Yoshikawa

The University of Tokyo
Department of Precision Machinery Engineering,
Faculty of Engineering,
The University of Tokyo,
Japan.

This paper deals with a method of constructing CAD systems and proposes a system developed by it. The general design theory is the methodology for construction, which covers the fundamentals of design. Techniques developed in the artificial intelligence are found useful for actual implementation of systems.

1. INTRODUCTION

THE STATUS OF SYNTHESIS AND DESIGN

First let us examine the nature of synthesis before going into subjects covered by the title of this paper. Making it simple, the model helps as shown in Figure 1. The main four blocks relate to a matter; they are Entity, Attribute, Function, and Value. They are the ways how people perceive a matter. Next is to classify the academic fields according to these four blocks.

Although Entity is an objective existence which is insignificant and depersonalized for human beings, people extract Attribute from Entity, then recognize its Function, and end up to appreciate Value. The line shows the process of signifying the natural existence of Entity from right to left. This process is called analysis. Those are separated as the following parts and each of them forms a special fields.

A1) From Entity to Attribute: It is a main work of science to extract Attributes from Entity, especially in the material research of physics. The usability of information given by Attributes is out of question in the case.

A2) From Attribute to Function: The concept of Function is formed when the usability of a matter is concerned. Function is the complex compound of Attributes. The field, which discovers Function from Attributes of Entity based upon the usability of a matter, is called engineering analysis.

A3) From Function to Value: The fields, which researchers the usability of Function quantitatively, is called economics. When it is applied to individuals, it is called microeconomics, and when it is applied to a society, it is called macroeconomics.

Let us think the convert way.

S1) From Attribute to Entity: When an Entity is created from Attributes, this process is generally called invention. Although it seems there is no process of generating an Entity satisfying Attributes without considering usability, we may define basic development, such as synthesizing materials of higher hardness than diamonds, because the process can be conducted without the consciousness of any usability.

S2) From Function to Attribute: Entity is formed to satisfy the given Function of the explicit concept of use. The acts of composing and compiling an Entity with representable Attributes after analysis, are called design.

S3) From Value to Function: In this case, forming a Function, which satisfies some Value is the process. Here, Entity itself is not dealt with. Rather practical value is created from the arrangement and composition of elemental Functions. Practical industrial administration and management technology are this type.

The current from left to right in the Figure shows the process of synthesis. One perceives the meaning of creating concreteness from abstracts in words like development, design, administration, and management. They are all synthetic. Especially it should be paid attention that the design in technology is composing an Attributes which satisfies given Function.

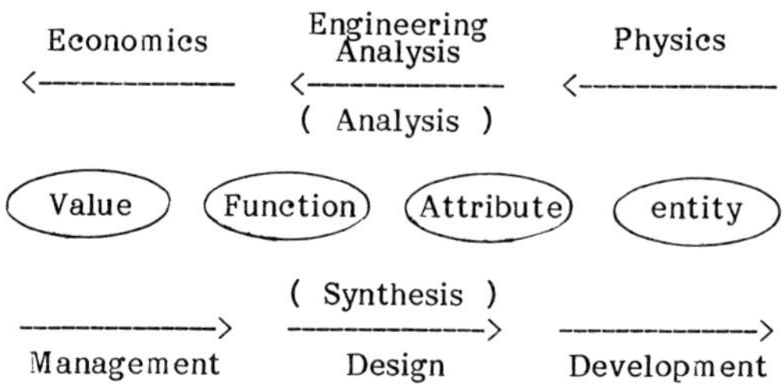

Fig.1 Fundamental Structure

Though it is hard to make an exact classification, Figure 2 shows the rough classification of each field of engineering according to Figure 1. There are fields researching specific features of Attribute and Function separately in regard to them.

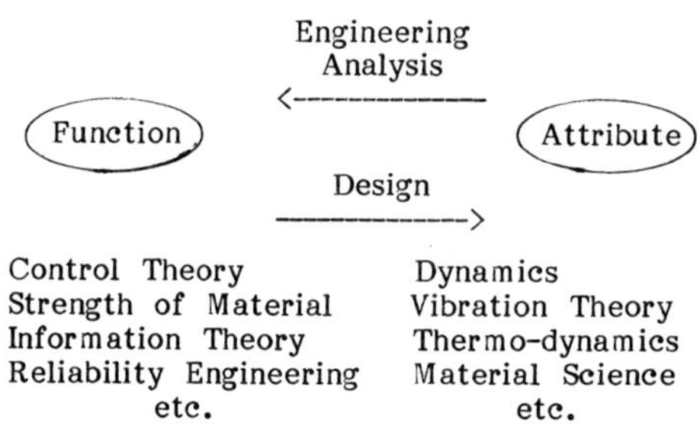

Fig.2 Relationship between Fields

Then the problem is the field researching the correspondence between Attributes and Functions. Given a machine, then extracting the distinctive feature of the machine from control-theoretical viewpoint is called the analysis of control system. The creation of mechanical system from abstract control system is called the synthesis of control system which is also included in the control theory. However, in reality, control theory emphasizes only analysis, and synthesis is a small part, or else it aims to make some parameters of the system suitable and not to make concrete systems. Thinking from the viewpoint of design in Figure 2, the field is extremely unfilled.

Comparing to other fields, control theory still emphasizes synthesis. Aiming the high practicability rate, the field of reliability engineering is created. Yet it emphasizes only analysis and the method of designing new systems with high reliability is taken only as examples.

The following is the main reason why each area concentrates upon analysis and not on synthesis. Under the control theory, a model of control system is made after extracting its meaningful Attribute of a machine or an equipment given. It makes possible to distinguish the features of control system exposed to different conditions and usages. In other words, it makes possible to obtain exerted Functions. The process of uncovering the functions by using a model is systematized and logical. This is the pivot point of control theory which is completely methodological as an analytic process of getting Functions from given Attribute, though it is not logical enough to construct models. The bases of this method are classical mathematics, such as differential equation and variation.

On the contrary, getting Attribute, which fulfill the given Function, is not systematized enough. For instance, if a control system of specific Function is given by a blockdiagram, it is a synthetic problem to find the system of Attributes, that is Entities, which is represented by this block diagram.

However, it is impossible to extract an attributive model from a block diagram which is functional model by a systematic method. Even if it is possible, creating Entity from such attributive models is hard. Substantially, there is no means to represent Attribute as a model which can be used in technological fields. So it is impossible to analogize the process from Function to Entity with the process from Entity to Function. Nevertheless, no special methods for synthesis exist. Consequently synthesis in reality relies heavily upon intuitions and experience.

2. GENERAL DESIGN THEORY [1]

2.1 Aims of the Theory

Designing is a typical intellectual activity which the human being performs. The skilled designers can design the highly sophisticated and complicated engineering systems such as turbines with their fertile knowledge about science and technology. The common people cannot design a turbine, but they design beautiful flower beds on their desks which will appear actually in their gardens next Sunday.

The knowledge which is particularly necessary for certain fields of industry is given by the school education and their own studies. But, on the contrary, the designing ability seems intrinsic for the human beings, and will be developed unconsciously by the repetitive experiences. The flower bed designer may not have consciously learned about the method of its design, though he must learn about the material of the flower bed, quality of soil, seeds and saplings of plants, fertilizer etc. The similar situation is found in the design of industrial products. The ability of design is thought to be improved only by experiences for years. A designer can acquire the necessary knowledge about the system he designs for a short time, but it is hardly possible for him to comprehend the method of design without long experience.

Such situation seems due to the fact that the process of design has been less understood. It is difficult to abstract the process itself from the practical designing activity even for designers with many experiences. We find some design theories, called engineering design, systematic design, design methodology etc, which insist that their methods are applicable for any type of design. But, unfortunately, they are not enough developed for the system construction of computer-aided design.

We have studied the general theory of design these several years.[2] It ultimately aims at clarifying the human ability of designing in a scientific way, and at the same time, producting the practical knowledge about the design methodology. It has been also revealed that the general theory of design is useful, or essential, to construct CAD system in a scientific manner. The theory indicates what CAD system should be, concerning the tasks to be imposed to computers, data structures, the role of a designer, design sequence, the interface of man and computer, etc.

2.2 Axioms

2.2.1 An Illustrative Example — Design of Smoked Meat

Let us consider a primitive world, where only three kinds of meat are found: fresh meat, putrid meat and dried meat (too much dried and leathery); all natural entities without any artificial processing. People of this world recognize and memorize them. Let S be a set, elements of which are natural entities:

s_1 = fresh meat

s_2 = putrid meat

s_3 = dried meat

Dogs and cats can also recognize and memorize them, but separately. On the contrary, human beings recognize and memorize them not only separately but also with classifications. They construct concepts about particular characteristics abstracted from these entities, giving classifications or structures upon them, that is, the set S. For example, let $T = \{ T_1, T_2 \}$ be a classification:

T_1 = (deteriorates with laps of time)

T_2 = (not eatable)

which have no one-to-one correspondence to each entity but to subsets of S. The subset generated by this classification has the semantic significance from the viewpoint of function or value of the entity.

Once the subset system T has been established, people can memorize the entities not only by their direct image but also by using the abstract character such as "not eatable and deteriorates with lapse of time" that is equivalent to the putrid meat in the present example. This means

$$\begin{aligned} s_1 &\in T_1 \cap T_2^c \\ s_2 &\in T_1 \cap T_2 \\ s_3 &\in T_1^c \cap T_2 \end{aligned} \qquad (1)$$

where c denotes the complement of the subset. It is assumed that people had the ability of logical operations such as And, Or or Complement besides memory, abstraction and classification.

After achieving those abstract concepts, people had started to think, memorize and communicate by using them instead of individual entities. Someday, a primitive man (or woman) thought accidentally about

$$T_1^c \cap T_2^c$$

which could occur without any difficulty if the combination (1) is familiar to him (her). This idea, however, must have been drastic and dreadful to the primitive because they had had a concept having a correspondence to a thing which did not exist in the realistic world. T_1^c T_2^c had no correspondence to a real entity but the value of it was higher than any existing entity. "Eatable and not deteriorate with lapse of time" had the highest value, and this conceptual combination was the <u>necessary condition</u> to invent the smoked meat which had been the first artificial entity for the human beings.

2.2.2 Formulation of Problem

2.2.2.1 Entity and its function

The necessary condition for designing, that is an act to create artificial things that have not existed in the real world, is the possible to form concepts of nonexistent things as the result of logical operation of knowledge about existent things. Therefore, we must prepare items necessary to discuss the designing, by defining the knowledge and the concepts that are to be subjected to logical operations.

It is popular to illustrate the designing process as shown in Figure 3, which is similar to a block diagram where the input is the requirement specification and the output is the design solution by which the product will be manufactured. Unfortunately, this illustration shows only a definition of design and any important information will not be derived from it. It is because definitions of the input and the output are not definite. On the contrary, in the field of control engineering for example, the description of this kind is very effective because the input and the output are definitely defined: both are time-series functions and thus the process is represented by the transfer function which has clear meaning mathematically.

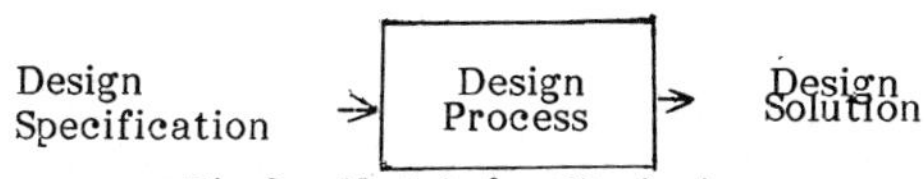

Fig.3 Model for Designing

The specification, the input of the designing process, is expressed by using languages, symbols, numbers and sometimes existent machines; and the design result, the output of the process, is described in drawings, models such as mock-ups and sometimes in languages. They are all more complicated than the time-series function. So, we must begin with the rigorous definitions of them.

DEF. 1 The entity set is a set that includes all entities in it as elements. By all entities, we mean entities that existed in the past, are existing presently and will exist in future. This set is denoted S'

DEF. 2 The item of attribute and its value. Attributes of entity are the properties; physical, chemical, mechanical, geometrical etc., which can be observed by scientific means. An entity has respective values for attribute items.

The number of attribute item may be infinite. But practically, we have only finite capacity of memory, as a result, we can give only the finite number of values of attribute for a given entity. It means that our recognition about an entity is restricted by some resolution power. But our real recognition can be treated as an approximation of the idealistic, that is infinite, case.

We shall use the following expression for the attribute. Let a_i be the item and v_{ij} be the value of a_i. Let us denote $a_i = v_{ij}$ when we give a value to the i-th attribute. Then,

$$\{ a_1 = v_{11}, a_2 = v_{21}, \ldots a_n = v_{n1}, \ldots \} = s_1' \quad S'$$

is a form which expresses an element s_1; of the entity set S'.

The next step is to give a rigorous definition upon the function, because the specification given to a designer is usually a functional requirement. Let us consider a pencil. The expression of attributes of a pencil is:

{ a_1(total form) = v_{11}(hexagonal shaft), a_2(tip form) = v_{21}(cone), a_3(structure) = v_{31}(black lead at the center of wooden shaft), a_4(length) = v_{41}(15cm), a_5(section) = v_{51}(hexagon with 4mm edge), a_6(color) = v_{61}(wood is brown, outside is green),

The way of selection of items is arbitrary, but the expression becomes the more accurate, the more the numbers of item are selected.

One of the functions of the pencil is that letters are written on paper with it. This function is exerted by some of the above attributes; a_1(total form), a_2(tip form), a_3(structure), a_4(length), a_5(section) in the present case. It is important that there are many, perhaps infinite number of, attributes which that no influence upon certain function: a_6(color) is the case in this example.

DEF. 3 When an entity is exposed to a circumstance, a particular behavior manifests corresponding to the circumstance. This behavior is called as visible function.

Different behaviors are observed for different circumstances. The total of these behaviors is called as latent function. Both are called as function inclusively. Sometimes, we use function for visible function when no confusion occurs. The field is the circumstance that is effective for a function to manifest.

Many different functions manifest on an entity according to fields.

fields \ attributes	a_1	a_2	a_3	a_4	...
f_1	v_{11}	o	v_{31}	v_{41}	...
f_2	v_{11}	v_{21}	o	o	...
f_2	v_{11}	v_{21}	v_{31}	o	...
.	...	...	...	...	...
.					

A function manifests as the result of activation of some number of attribute, while others are latent, denoted zero in this table.

2.2.2.2 Function as a concept

As mentioned in the preceding item, the ability of concept formation and of its logical operation is the necessary condition for human being to design. So, here, let us give some definitions about concepts.

DEF. 4 Concept of entity is a concept that one has formed according to an actual experience of an entity. In this concept, any attributes or functions are not abstracted of the entity. The entity is totally a concept for one.

Sometimes, we have such experience that an image appears suddenly in our memory of an impressive happening, though we have no analytical data about the origin of the image. This is due to such total memory, which is defined as concept of entity.

DEF. 5 Abstract concept is derived by the classification of concepts of entity according to the meaning or the value of entity. As a result of classification, we get classes, each of which includes entities carrying the common meaning or value; and corresponds to a particular abstract concept.

It is not easy to give concrete names to those classes. It could be the problem of concept formation which is taken up in the fields of psychology, artificial intelligence, and sometimes philosophy. But here, we do not go into the detail of this problem only considering the abstract definition which will not contradict the results of those fields. We give narrower definition of abstract concepts which are effective for further discussion of the design problem.

DEF. 6 Concept of attribute is a kind of the abstract concept. This concept is effective for one to recognize entities. This concept has no direct relationship to value. The classification due to this concept is used when one intends to classify entities independently of values, that is scientifically.

DEF. 7 Concept of morphology is sub-class of the concept of attribute. When attention is given intensively to morphology, this concept forms.

DEF. 8 Concept of function is a sub-class of the abstract concept. When attention is given to the functional value, it forms. In other words, when an entity has a peculiar latent function and is able to turn it into the visible one by an existent field, then the entity is classified into a class which was made under the name of the particular function.

These concepts defined above also make up sets. Following these definitions, let T, T^0, T^1 and T^2 be sets of the abstract concept, the concept of attribute, the concept of function and the concept of morphology respectively.

2.2.3 Systems of Axioms

In this chapter, some axioms are given. The reason why we call them axioms is that they seem not to contradict to our intuition and our general experience but are difficult to be proved directly and to be verified by experiments; and are powerful for deducing many useful theorems which are opened to the experimental verification and the counter-evidence. The goodness of axioms is measured by the number of useful theorems deduced from them. The following axioms are concerned with the concept of human beings.

AXIOM 1 (Axiom of recognition) Any entity can be recognized or described by the attributes.

AXIOM 2 (Axiom of correspondence) The entity set S' and the set of concept of entity (ideal) S have one-to-one correspondence.

This axiom is for ideal knowledge which has infinite capacity. Unfortunately, our real knowledge is far from the ideal, but the former will be defined as one deviated from the latter. This definition is efficient for the discussion of the real knowledge when the process of design is studied in coming chapters. It is similar to the crystal physics. Real crystal is understood as of deviation due to structural defects from the ideal crystal structure.

Function space (S,T^1) Attribute-concept space (S,T^2)

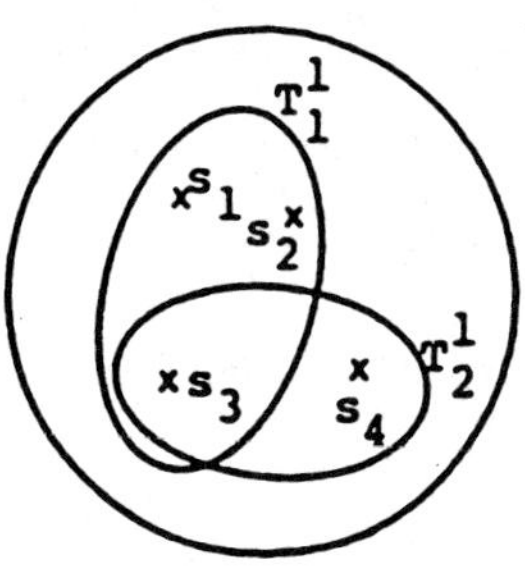

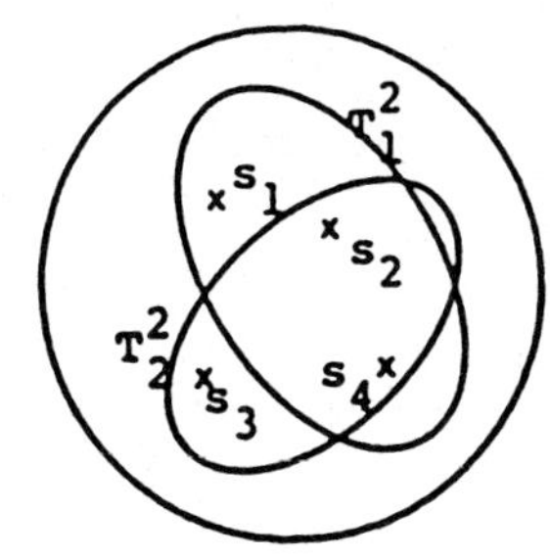

Fig.4 Topologies in concept set. Entity set S = { s_1 = crayon, s_2 = pencil, s_3 = brush, s_4 = electric pen } is given structures by the functional topology T^1 = { T_1^1 = "to write letters on paper", T_2^1 = "strong against shock" } and the attribute-concept topology T^2 = { T_1^2 = "having a sharp tip", T_2^2 = "having a long shaft" }

AXIOM 3 (Axiom of operation) The set of abstract concept is a topology of the set of entity concept.

This axiom insists on the following properties.

a 1) All of the entity concept, that is the universe set of the entity concept is an abstract concept. Inversely, null set of entity concept is also an abstract concept.
a 2) By the definition, an abstract concept is a subset of the entity-concept set. The intersection of the subset and that of another abstract concept is also an abstract concept.
a 3) A union of abstract concepts is also an abstract concept.

These properties are expressed in terms of mathematical words as follows:

b 1) $\phi \in T, \quad S \in T$

b 2) $T_1 \in T, \quad T_2 \in T \quad \rightarrow \quad T_1 \cap T_2 \in T$

b 3) $\bigcup_{\lambda \in \Lambda} T_\lambda \in T$

Hence, T being the topology of S. This axiom is concerned with the static structure of knowledge or memory of human being. But also it includes the intersection and union which are the dynamic operation of human thought; and thus very important for designing problems.

An example is illustrated in Figure 4. This axiom seems self-evident intuitively, but cannot be verified directly. Experiments on concept formation or thought process will provide it with the evidence in future.

2.3 Theorems

2.3.1 Principles for Deduction of Theorems

Axioms concerned with the concept are proposed in the preceding chapter, on a line such that they do not contradict the knowledge obtained in other fields and we can accept them intuitively. The subject of the present chapter is to deduce theorems from them. Before stating the individual theorems, the fundamental standpoint by which theorems are deduced is revealed here.

The central aim of the deduction is the clarification of designing process. And practically, we try to find out the necessary condition for designability, to supply the standard of the design strategy and to extract algorithmic part in designing etc.; which will be useful for constructing rational computer aided design systems. For these practical purposes, some conditions that occur that actual circumstances such as economy and workers ability must be considered. It means that the deduction must sometimes refer to those practical conditions deviating from pure theories and logics. We introduce special assumptions and create necessary limitations in such cases. But in the present chapter, we shall not take up those practical aspects but confine the discussion into the theoretical problems which will be the fundamentals for those practical purposes. The practical problems, especially concerned with actual designer's designing process deviated from the ideality, will be discussed in the following chapter.

2.3.2 Theorems about Requirement Specification

The knowledge consists of the entity concept and the abstract concept. The following defines the ideal knowledge.

DEF. 9 Let the ideal knowledge be that knows all of the elements of the entity set and can describe each element by abstract concepts without ambiguity.

The Real knowledge intends to approach toward this ideal knowledge. We always acquire incremental knowledge by increasing the number of entity concept which gives the structure to the former, in order to memorize or describe the entity as accurate as possible. Such tendency makes the real knowledge closer to the ideal one.

THEOREM 1 The ideal knowledge is the Hausdorff's space.

Proof In the ideal knowledge, the set of entity concept S is identical to the set of entity S'. The subset system T is introduced into S. Let s_1 and s_2 be the elements of S. The neighborhood systems of them are described as $T(s_1)$ and $T(s_2)$. The necessary condition to describe any entity without ambiguity by the abstract concepts is that two entities, s_1 and s_2, are discriminated by the abstract concepts. Thus, there must be such $T_1 \in T(s_1)$ and $T_2 \in T(s_2)$ where

$$T_1 \cap T_2 = \phi$$

holds. Conversely, suppose that a space satisfies the Hausdorff's condition, then the uniqueness of convergence of point series is guaranteed, hence the unique entity by the abstract concepts.

DEF. 10 The requirement specification T_r is that designates the function of the designed machine (system) by using the abstract concepts. Thus, the specification $T_r \in T$ and $s \in T_r$ is the design solution. Conversely, suppose that s is given, then the specification of s' is any of the neighborhood system of s, that is, any $T(s) \in \{ T(s) \}$. Practically, the specification is described in terms of functions, hence $T_r \in T^1$ which is called a functional specification.

THEOREM 2 The specification can be described by the intersection of proper abstract concepts.

Proof This result is derived from AXIOM 3. For any T_r, we can select the set of T_λ so that $T_r = \bigcap_{\lambda \in \Lambda} T_\lambda$.

Here it is pointed out that the specification can be classified according to its quality. Three classes are as follows:

i) $T_r = \{ s_1 \}$

ii) $T_r = \{ s_1, s_2, \ldots \}$

iii) $T_r = \{ \phi \}$

They are called as an ideal specification, an ambiguous specification and a contradictory specification respectively.

THEOREM 3 The design specification, unless contradictory, is a filter.

Proof The specification which is not contradictory is the subject in this theorem. Consider $T_1, T_2 \in T$. Let T_1 be a specification, and if $T_1 \subset T_2$, then T_2 is also a specification. If T_1 and T_2 are specifications for the same solution, then $T_1 \cap T_2$ is also a specification. $T_1, T_2 \neq \phi$. A subset T_j with such properties is called an element filter. Thus, the set of specifications is a filter as a subset system.

THEOREM 4 From the preceding definition, the following properties are easily deduced.

i) $T \supset T^0 \supset T^2$

ii) $T \supset T^1$

But, the relationships between T^0 and T^1; and between T^1 and T^2 are not definite, hence discussed later.

From the practical point of view, it is known of primary importance in the designing process how the specification is described and detailed. The necessary labour for this process be comparable to that necessary for seeking the design solution after receiving the complete specification. Therefore, it is essential to clarify the substance of specification. This can be done by investigating the function space which is a model of human concept system. It is noted that the language will take important role in the investigation.

2.3.3 Theorems about Design Solution

DEF. 11 The design solution is an entity concept that is included in the correspondent specification T_r and carries all necessary information for manufacturing it.

The following theorem is deduced easily.

THEOREM 5 The entity concept in the ideal knowledge is a design solution.

Proof In the ideal knowledge, once an entity concept is selected all the abstract concepts that are the neighborhood of the entity concept are known immediately, and among them necessary information for manufacturing it is found.

THEOREM 6 The entity concept in the attributive space (S, T^0) is a design solution.

Proof This theorem is based upon the fact that the manufacturing means for an entity are solely dependent upon its attributes.

THEOREM 7 The design solution is represented by the intersection of properly selected elements (subsets of S) of the set of attribute concept.

Proof This theorem insists that $T = \bigcap_{\lambda \in \Lambda} T_\lambda$, when $T_\lambda \in T^0$ is a solution. It is evident when only one entity is included in T, that is, $T = \{ s_1 \}$, because necessary manufacturing means are found in the neighborhood system of s_1. But suppose that $T = \{ s_1, s_2, \ldots \}$, then subsets selected from neighborhood system of T will give manufacturing means but they are not unique, hence multiple solutions. Therefore, in case where T includes multiple entity concepts, the solution T is thought as an approximation. On the contrary, when $T = \{ \phi \}$, the solution is contradictory and no solution is existent.

2.3.4 Theorems about Designing Process

In the preceding chapters, it was revealed that the specification is described by the combination of abstract concepts and the solution is described by that of attributive concepts. So the rigorous definition of designing is given as follows:

DEF. 12 Designing is to designate a domain on the attributive concept space (S, T^0) that is included by that on the abstract concept space (S, T).

THEOREM 8 If the abstract concept space on which the specification is described is limited within the attributive space, as the subspace of the former, then the designing is completed when the specification is described.

Proof This is evident because the specification is identical to the solution in this case. The extreme example is that one orders a machine by pointing it on a catalogue.

THEOREM 9 In the ideal knowledge, the solution is obtained immediately after the specification is described.

Proof The specification is $T_r = \bigcap_{\lambda \in \Lambda} T_\lambda$ according to THEOREM 2. If λ increases, T_r decreases. $T_r \neq \phi$ according to THEOREM 3. Because the space satisfies the Hausdorff's condition, T_r converges to a point s_k as λ increases. In the ideal knowledge, any neighborhood of s_k is known of the attributive space, hence s_k being the solution.

Suppose that the knowledge is not necessarily ideal, and the specification is described in terms of functional concept. Then it is the necessary and sufficient condition for the possibility of design space and the attribute space. From this we add another definition for design:

DEF. 13 Designing is a mapping of a point in the functional space onto a point in the attributive space.

When the design task is limited to a narrow field; for example, in the designing of a beam subjected to some amount of load, designers know a calculating formula of beam dimensions according to the applied load. This formula can be said a known mapping which gives the solution, the beam dimensions in this case, correspondent to the specification, the load, by an algorithmic way. We call such case an algorithmic knowledge which has the similar performance as the ideal one. The automation of design will be improved along two lines: aiming ultimately at the ideal knowledge and the algorithmic knowledge. They are the increase of function-attribute correspondence knowledge by using bigger memory capacity and the preparation of more calculating formula by developing the mathematical models of function-attribute correspondence, respectively.

THEOREM 10 If design is possible, the identity mapping from the attributive space (S, T^0) to the function space (S, T^1) is continuous.

Proof This theorem insists that it is the necessary condition of designability (possibility to design) that the topology of the attributive space is stronger than that of the functional space, that is,

$T^0 \supset T^1$. Suppose that $T^0 \supset T^1$ does not hold. Then, by describing a specification by $T^1_1 \in T^1$, it is possible to occur that T^0_1 is not existent in the attributive space such that,

$$s_1 \in T^0_1 \longrightarrow s_1 \in T^1_1$$

which is the necessary condition for T_1^0 to be a solution. In other words, in the design to find an expression of s_1 in terms of attribute T_1^0 by giving an expression of s_1 in terms of function T_1^1, it is possible to occur:

$$\exists s_1, \ \forall T_1^0 \ (s_1 \in T_1^0 \text{ and } s_1 \notin T_1^1).$$

Then it is probable that the solution T_1^0 permits the entity s_1 which does not satisfy the functional specification T_1^1. Thus unless $T^0 \supset T^1$, the designability is not guaranteed.

THEOREM 11 If two design solutions can be discriminated functionally, then $T^0 \subset T^1$.

Proof Similar to the proof of the Theorem 10.

2.4 Real Process by Designer

2.4.1 Imperfections in Real Knowledge

In this chapter, we discuss about the real designing process by designers. As mentioned in the last chapter, under the assumption of the ideality of knowledge structure some theorems were deduced which predict the possibility of design in systematic ways. Unfortunately, this is not the case for real knowledge which designers have. It is necessary, then, to find out some realistic operation that enable them design in spite of their imperfections. For this purpose, these imperfections must be clarified. The ideality was based upon the infiniteness of memory capacity and operational speed. On the contrary, the real knowledge is peculiar for its finiteness of them. Due to this finiteness, some imperfections occur as follows:

1) Imperfections concerned with the entity concept: it is possible that some abstract concept includes no entity concept ("vacancy").

2) Imperfections concerned with the categorization of entities by the abstract concept: the topology is not perfect, hence insufficient separation of entities ("dislocation"). Some entity can be isolated without categorization ("impurity").

3) Operational imperfections: due to erratic set-theoretical operation, it is possible that the filter condition, Theorem 3, is violated in the detailing process of specification ; or else, it takes infinite time for the convergence.

4) Imperfections concerned with the mapping: it is impossible to expect a known correspondence between different topologies.

5) Imperfections concerned with the analysis: in the designing process analytic operations are indispensable as shown in Theorem 9. The solution cannot be sometime described even when we succeed to converge into a known entity.

These are shown in Figure 5 schematically.

The designers, who cannot avoid these imperfections, still succeed in production of excellent designs overcoming them skillfully. In order to understand the designer's designing process, it is essential to find out the necessary conditions for the possibility of design in spite of these imperfections.

Generally, it is concluded that some operational rules and restrictions are necessary for designer's design besides those fundamental characteristics of knowledge: the entity concept, the abstract concept and their set-theoretical relations. This is mentioned in the following theorems.

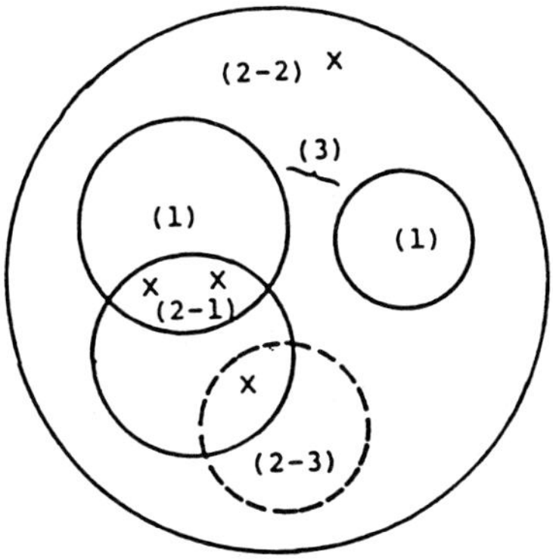

(1) No entity concept (vacancy)
(2-1) Insufficient separation of entities (dislocation)
(2-2) Isolated entity concept (segregation)
(2-3) Ill-classified entity concept (impurity)
(3) Violation of filter condition in specifications (grain boundary)

Fig.5 Structural imperfections in real knowledge

THEOREM 12 The design process appeared in Theorem 9: the convergence to an entity in the abstract concept space and the discovery of a necessary filter in the attribute space; does not operate in the real knowledge.

This leads to the following theorem.

THEOREM 13 In the real knowledge, the design is possible if and only if there is some any rule of direct correspondence between the topologies of abstract concept and of attributive concept without intervention of the entity concept.

We must find out some realistic, that is approximate, correspondences between topologies of functional concept and of attributive concept under the restriction described in the Theorem 13 which insists that only such direct correspondence gives the real designer the possibility to design. First, we consider about the correspondence in the Euclidean space. Assume that both the functional space and attributive space are Euclidean. Then, the correspondence can be described as a mathematical equation. The equation will be solved rigorously or the solution will be given as the limit of a numerical series. As the analogy of this Euclidean case, let us consider that there are two cases in the general topological space too.

(A) The explicit correspondence is known.

(B) The method is known to find the directed series of points which will converge into the solution.

The former is called correspondence type and the latter is called the convergence type.

2.4.2 Some Models

In this section, some models will be proposed that assure the possibility of design by real designers under the restriction of the Theorem 13. They are classified into two categories, the correspondence type and the convergence type, mentioned above.

(A) Correspondence type

(1) One-to-one correspondence model (Catalogue model)

When we assume that the cardinal number of both the functional space and the attributive space are of finite number, then we obtain the finite boolean algebra which represents requirement specifications and solutions. In this case, it is possible that the each specification in the functional space will be given a solution with the explicit correspondence, as shown in Figure 6. In this model, the design problem is reduced to the data retrieval problem. It means that no artifact will be created, which has not existed in the past, by this model.

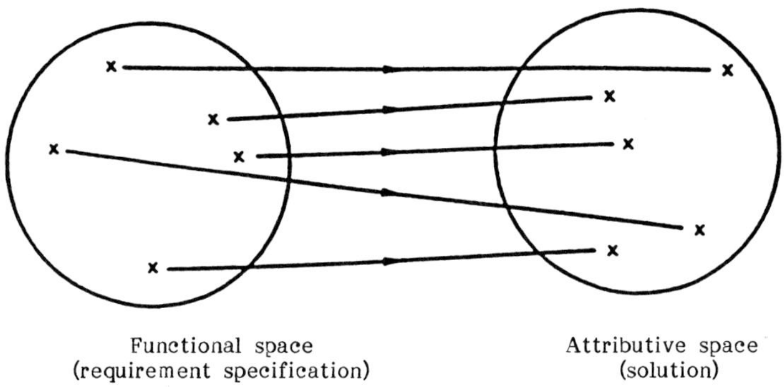

Fig.6 One-to-one correspondence model

(2) Calculational model

If a designer is allowed to make such approximations, firstly that the numbers of interesting items of function and of attribute are finite, and secondly that the value of these items are represented by number, then he will be able to reduce the design problem to the mapping problem between two Euclidean spaces. By selection of the items:

$$\text{Attribute} : T_1^0 , T_2^0 , \ldots , T_n^0 \in T^0$$

$$\text{Function} : T_1^1 , T_2^1 , \ldots , T_m^1 \in T^1$$

The following values (vectors) will be obtained.

$$\text{Attribute} : Y = (y_1, y_2, \ldots, y_n)$$

$$\text{Function} : X = (x_1, x_2, \ldots, x_m)$$

The Theorem 10, which gives the necessary condition for design, imposes $n \geq m$ in the present case. From the Definition 3, it is easily understood that there is such relationship as following.

$$X = f(Y)$$

The solution exists if f^{-1} exists. Usually the above equations include inequalities, thus the solution of it will be shown as some region, as shown in Figure 7.

Spec. $X_r(x_1, x_2, \dots)$

Relation $X = f(Y)$

Surfaces

(1) $x_1 = f_1(Y)$
(2) $x_2 = f_2(Y)$

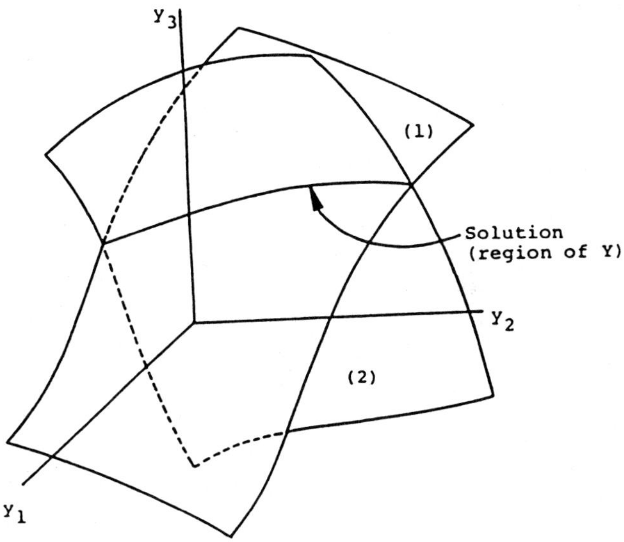

Fig.7 Calculation model

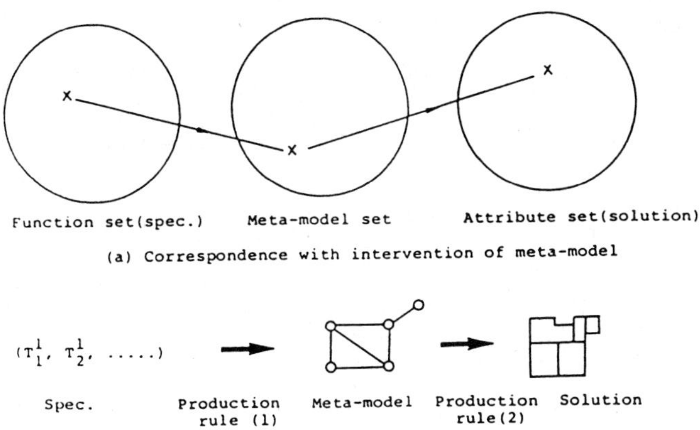

Fig.8 Production Model

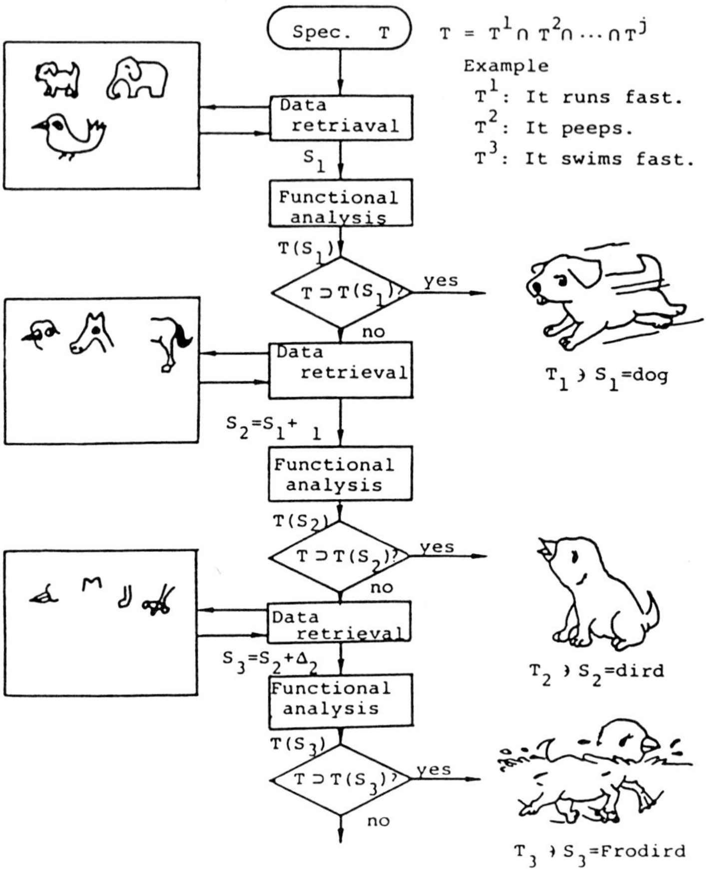

Fig.9 Paradigm model

(3) Production model

If there is a set of rules which produce a solution in terms of attribute by giving a specification in terms of function, we can say that the design is algorithmic with production rules. The calculational model is one of the production model. But, here, we consider the production rules as more general than numerical equations. As mentioned in the Definition 13, there will be possibly an algorithmically ideal knowledge by which any design is possible. In the knowledge, however, it is only possible when some restriction is imposed on the object of designing. For example, if we consider the mechanical machines, the meta-model[3] of it will be useful as an intermediate medium of the production rule. As shown in Figure 8, the functional specification will be transformed to a graph-theoretical representation, that is a meta-model, of a machine to be designed, and then the model will be transformed to a machine; by using production rules respectively.

(B) Convergence type

(4) Paradigm model

Given a specification $T = T_1 \cap T_2 \cap \ldots \cap T_n$. Assume that a solution s_1 is proposed, which satisfies

$$s_1 \in T(s_1) = T_1 \cap T_2 \cap \ldots \cap T_k \cap \overline{T}_{k+1} \cap \overline{T}_{k+2} \cap \ldots \cap \overline{T}_n.$$

Therefore, s_1 satisfies only a part of specification but not all. Then assume the next proposal s_2

$$s_2 \in T(s_2) = T_1 \cap T_2 \cap \ldots \cap T_k \cap T_{k+1} \cap \overline{T}_{k+2} \cap \ldots \cap \overline{T}_n.$$

satisfies a part again, but improved in comparison with s_1. Then, the move from s_1 to s_2 satisfies the property of directed series of points. Let us call these points, which are the proposed solutions, as paradigms. If there is a rule which produces a directed series, we can say that the new proposal will be better than the older. Then, we must develop a policy to produce the directed series actually. Firstly, we hypothesize the following properties.

Hypothesis 1 (or Axiom 4) The more similarities of function there are between two entities, there are the more similarities of attribute between them. The converges also holds.

Hypothesis 2 The entity concept in the designer has a hierarchical structure.

On these assumptions, a designer can propose a preliminary solution that could satisfy the specification given. Usually, however, it only partially satisfy the specification. The similarity of the attributes is measured by the extent of community of elements included in them. Therefore, the designer will try to change the irrelevant element of attribute, in order to improve the similarity of function toward the functional specification, that is the requirement.

An example is shown in Figure 9.

3. PROSCODE : A CAD FOR CONCEPTUAL DESIGN

3.1. Detail Process by Real Designer

In the superman's process, a designer can add another function so as to generate a more detailed requirement without violating the filter condition mentioned in Theorem 3. And he can assign an entity that satisfies the requirement at some stage of the detailing. When the domain of intersection in the functional space becomes enough small, he can declare that one of the entities included in the domain is a solution. In order to realize a process which approximates the above one, it is necessary to investigate the property of designer's knowledge.

Let us consider a simple example.

<u>Original requirement</u>: A machine that transforms revolutions safely among people

The superman can judge immediately that this requirement satisfies the filter condition. If so, he will select an entity concept that satisfies the requirement. Then, he will analyze the concept to enumerate every attribute of it. If he selects such a machine as a gear train which is completely covered by a housing, then he will <u>measure</u> the dimensions, tolerances,'surface finish, material, strength etc. of the machine selected from his memory.

When he is not a superman, the process will be as follows. If he is lucky, he may know a gear train which is completely covered by a housing as a machine that satisfies the original requirement, then he behaves himself just as a superman. However, If he is not so lucky, he may not know it. Then, he starts with survey of his knowledge to translate the description of original requirement. He may get a new description.

translation 1: A machine that transforms revolutions, and dangerous part of which is isolated from people.

In this case, he applied his following knowledge.

knowledge 1: A machine is safe if the dangerous part is isolated from people.

This translation is proved that it is successful afterwards, because he can proceed further by another translation.

translation 2: A machine that transforms revolutions and totally covered,

where he applied

knowledge 2: If a machine is totally covered it is isolated.

As for the transforming function, he will think of two revolving shafts as a direct conclusion of the original requirement. But he needs a knowledge to proceed design steps further. When he has

knowledge 3: If two gears are coupled, the transformation of revolution will be obtained,

translation 3: A machine which has two shafts on which gears are attached respectively and couple, and totally covered.

Although he is not a superman, he may have enough knowledge about shafts, gears and covers, because they are fundamental mechanical elements of limited numbers. At this stage of design, he can draw a structural image of a machine to be designed. Following knowledge.

knowledge 4: When a stationary part supports a rotational shaft, the friction between them must be low.
knowledge 4': When a part is fixed to a rotational shaft, the friction between them must be high.
And,
knowledge 5: The ball bearing has low friction.
knowledge 5': The key products high friction.

will produce more concrete machine design.

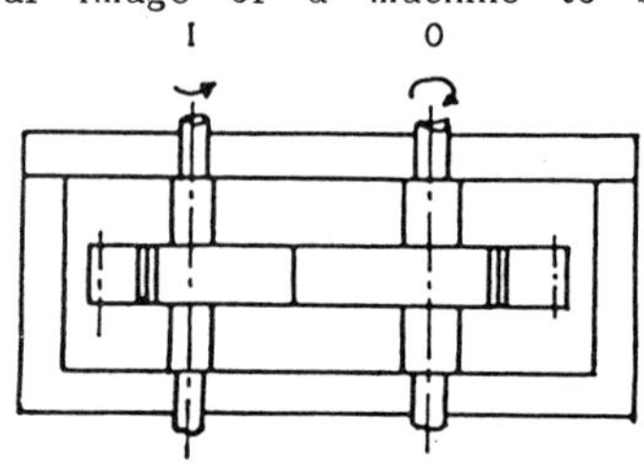

Figure 10 Projection of Gear Box.

The above mentioned example of real designer's process reveals that he replaces the direct assignment of an entity, which is observed in superman's process, with stepwise transformation of requirements (function) into solution (entity). Each step of the transformation is realized by his pieces of knowledge he achieved in the past. The translations he made are categorized as follows.

type 1: function -> attribute
ex. safe -> isolated, transform revolutions -> two related revolutions.
type 2: attribute -> entity
ex. isolated -> cover, two related revolutions -> gear pair, revolution -> shaft,
type 3: connection of entities function
ex. shaft and housing -> low friction; shaft and gear -> high friction.

In the general design theory, these translations are explained as follows.

As mentioned in chapter 2, the design is a mapping from the functional space to the attributive space. Because a real designer can not find out an entity, that satisfies the original requirement, in the attributive space only by the converging process in the functional space, he tries to find out relevant replacement of function with attribute. That is to find out T_i^0 in the attributive space such as

$$T_1^0 \subset T_1^1 , \quad T_2^0 \subset T_2^1 , \ldots$$

where $T_R^1 = T_1^1 \cap T_2^1 \cap \ldots$ is the original requirement in the functional space. If we find such T_i^0 , it is easily proved that

$$T_R^0 \subset T_R^1$$

where $T_R^0 = T_1^0 \cap T_2^0 \cap \ldots$ This is illustrated in Figure 11, showing that if an entity is selected as a solution from T_R^0 , it must be included in T_R^1 , which is the original requirement, hence a satisfactory solution. The above-mentioned type 1 translation corresponds to this matter.

The translation of type 2 is the assignment of an entity which is included in T_R^0 . This seems easier than the direct assignment in T_R^1 . But, in case where the functional requirement is fundamental, such as to fix two bodies, to rotate a shaft with low friction, one can assign an entity directly from T_R^0 . Therefore, we obtain a translation "function -> entity" as a reduced form of "function -> attribute", "attribute -> entity".

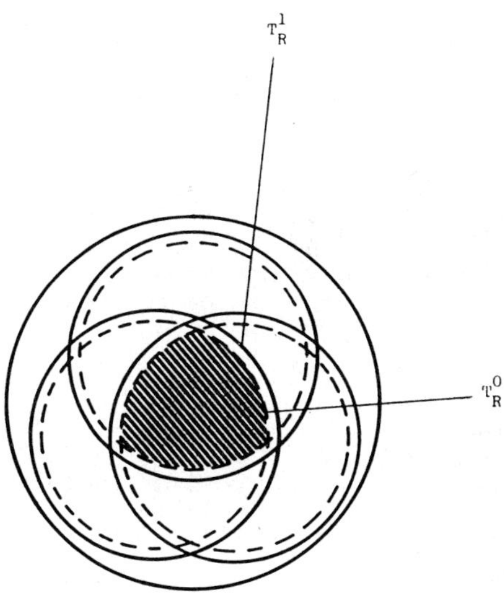

Fig.11 Requirement in Function Space T'_r and That in Attributive Space

Type-3 translation comes from such characteristics of function that is composed of several components of sub-function. Each sub-function is exerted by a respective sub-system of a whole machine. Usually, a new requirement appears between two sub-systems, which will be satisfied by a partial design. This problem has not yet been theoretically dealt with in the general design theory, but it is a very important one because the structure of function or the operation of function may be reflected in it. In other words, the new requirement is concerned with the necessary conditions for a machine to maintain its original function. Some necessary conditions were extracted by machine anatomy.

3.2 Design and Knowledge Engineering

The general design theory aims at establishing a mathematically rigorous model of designing process. The theory, however, does not reveal every designing process of designer precisely at the present stage. It is possible, at the moment, to construct a realistic model of a designer on the theory, that is, on the set-theoretical model of the human knowledge. The example of the last chapter showed that a model of a real designer was extracted from an empirical observation.

It is readily seen that those translations in the last chapter bear a resemblance to production rules in knowledge engineering. But the general design theory requires some characteristic natures on the rules for guaranteeing the possibility of design. In other words, the theory guides to fabricate rules which are effective. Those natures are realized as shown in the following examples.

To type-1 translation, we assign such a production rule.

type-1 rule: IF attribute, THEN function.

The nature of this rule must be

$$\bigcap_j T_j^0 \subset \bigcap_j T_j .$$

Example 1. IF adhere to paper by pressing (T_1^0), and have visible color (T_2^0); THEN write letters on paper (T_1^1).

When we neglect T, we may violate the nature mentioned above, resulting false design such as a pencil with colorless lead. It must be noted that the application of this rule guarantees to design something which we can write letters on paper, but it is also possible to overlook other solutions such as ink with metallic tip. In order to prevent it, we need another rule.

Example 2. IF be enough strong (T_1^0), and keep material (T_2^0) which be liquidus (T_{21}^0), and have visible color (T_{22}^0); THEN write letters on paper (T_1^1).

This rule also does not violate the requested nature. This fact reveals that we have several solutions which fulfill the vague functional requirement. They are all relevant solutions. But when we give more detailed functional requirements, some of them will be ruled out.

The production rule for the type-2 translation is:

type-2 rule: IF entity, THEN attribute.

The similar nature is essential here, such as

$$\{ S_j \} \in \bigcap_j T_j^0$$

Example 3. IF black lead (S_1), THEN adhere to paper by pressing (T_1^0), and have visible color (T_2^0).

For the type-3 translation, we give the following production rule.

type-3 rule: IF entity and entity, THEN function.

Example 4. IF black lead (S_1) and wooden frame (S_2), THEN high friction.

This type of rule generates new requirements for local design, which assures the incremental steps of design, that is concretization.

In the designing process, it is normal that the original requirement is given in the functional description. Then the system will look for a THEN clause of type-1 rules, firing an appropriate rule. Then, IF clause of the fired rule will be connected to the THEN clause of the type-2 rule. Thus, we shall find a network of entities as a solution. But in order to assemble these entities physically, the connecting parts must be designed in more detail. Type-3 rules give us new requirements for this local design under the restriction necessary to maintain the original function. We obtain functional requirements from those rules, so we return to type-1 rules to continue the design process.

3.3. A Pilot System PROSCODE-1

Tomiyama of our laboratory has recently implemented a pilot system PROSCODE-1 (PROduction rule System for COnceptual DEsign). This system works for machine design. Fundamentally, the knowledge of the system is categorized into two groups:

(1) Entity (S) has function (T^1).
(2) Function (T^1) is required for connection of two entities (S_1, S_2).

The format of the rule is slightly different from the explanatory examples mentioned in section 3.2.

At present, we have not completed a method to describe functions that can be applied for computer implementation. So, we represent functions of machine mainly by power transformation. There are some additional functions not represented by power transformation which commonly appear in mechanical systems. Then we obtain two kinds of rule for the first category.

R1: IF (entity (entity for input + entity for output)),
THEN (function represented by power transformation).
Example. IF (motor (electric line + shaft)),

THEN (electric -> mechanical rotation).

R2: IF (entity),
THEN (function)
Example. IF (bolt-nut),
THEN(fix).

For category 2, the rule is as follows.

R3: IF (entity + entity), THEN (function)
Example. IF (shaft + housing), THEN (shaft support).

Another type of rule is supplemented.
R4: IF (assembly), THEN (entity + entity)
Example. IF (bolt - nut assembly), THEN (bolt + nut).

In order to investigate the fundamental capacity of the present system, 21 rules were implemented and the functional requirement "to transform electric power into hydraulic power" is given. Then, the result shown in Figure 12 was obtained.

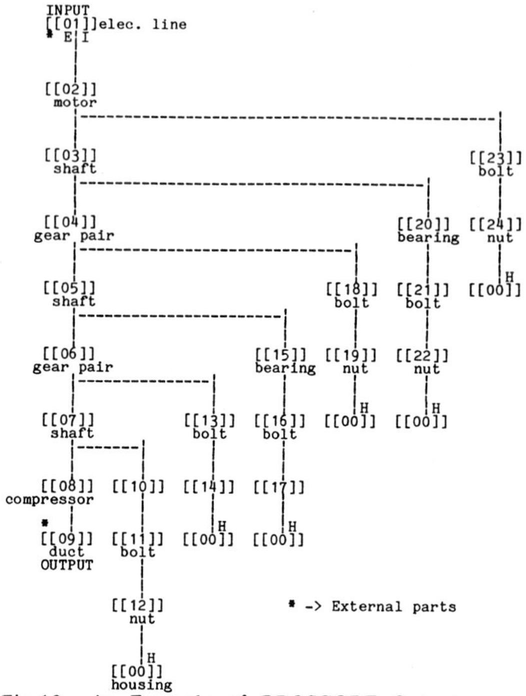

Fig.12 An Example of PROSCODE Output

4. FRODIRD: A CAD FOR DETAILED DESIGN [6]

4.1 Frame Operating System

According to General Design Theory [1], design is a process that makes directed series of points in the attributive space and converges them into a final solution. This process can be well expressed if we use frames to represent design objects

(Figure 13).

A design object is represented by a set of attributes, and an attribute is described as a combination of its name and value. But, actually, each attribute is accompanied by structural and physical restrictions that should be satisfied to ensure that the object may exist as an entity. Accordingly, data structures that can represent not only attributes of objects but also these restrictions are necessary.

Most of the operations that help to advance the design process are heuristic and peculiar to design objects. So, if plain knowledge structures are used to represent them, some representations, such as conditions of the application, become complex.

We can expect some phases of growth of CAD system using frames:

(1) the system that can in some way replay the behavior of designers;
(2) the system that can verify the realization of design objects;
(3) the system that provides general design methods by means of reorganization and integration of design knowledge.

All knowledge used in this system is represented by frame type data structures. One frame is a set of slots, and each slot has its name and values. There are two types of frames: Object-frames and Action-frames.

Design objects are prescribed by some attributes, and each attribute is represented by a set of item names and their values. Therefore, in Object-frames, which are used for representing design objects, names of attributes correspond to slot names and their values to slot values.

As a matter of fact, each value of an attribute, that prescribes a particular aspect of the design object, is subjected to structural and physical restrictions in order to exist as an entity. For frame operations to make sense with regard to design, it is indispensable to describe these restrictions. Therefore, such restrictions and/or design knowledge are added to slots in the form of procedures; they are called Attachment.

On the other hand, in order to express design action, a frame just like a script frame is required. We realized Action-frames as script frames. An Action-frame is evoked by giving its name and some of its arguments, and then the Action Text, defined in its slots, is executed. In this Action Text, the operation sequence of design as well as selection and evaluation of the candidate design solution are described according to the notation of the frame operating language (Figure 14).

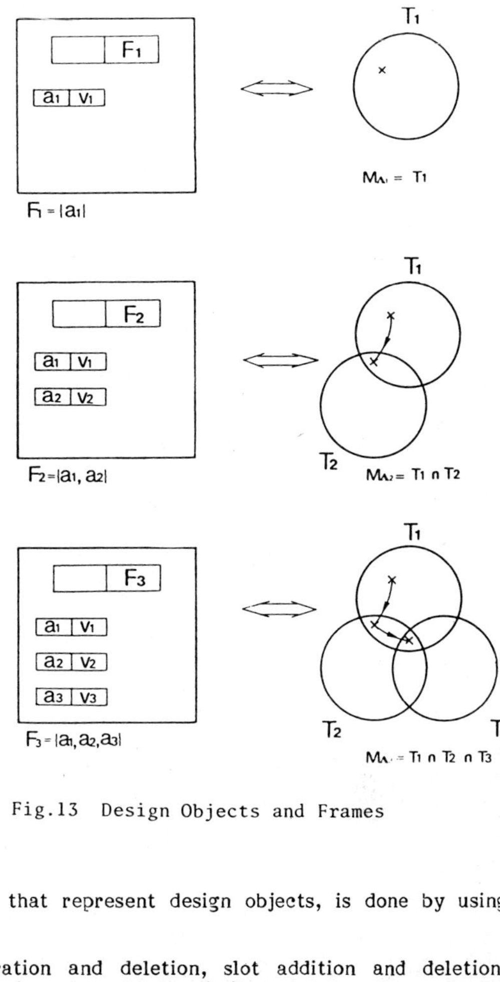

Fig.13 Design Objects and Frames

The actual operation on Object-frames, that represent design objects, is done by using

Basic Operations such as frame generation and deletion, slot addition and deletion, filling slots with values and checking of various kinds of flags in the slot. Design procedures in Attachment and Action Text are described in combination with these Basic Operations.

There is another classification of frames: Base-frames which represent abstract concepts and Instance-frames which represent concrete concepts. Here, the design objects are represented as an assemblage of Instance-frames, and each Instance-frame is generated through instantiation of its superior Base-frame. In the case of Action-frames, an Instance-frame is created every time a Base-frame is evoked. Frame operations, as mentioned above, are illustrated in Figure 15. Each Instance-frame is generated by instantiating a Base-frame of the upper level.

```
   <Action_Text> ::= <Action_Element> ¦ <Action_Element><And><Action_Text>
<Action_Element> ::= <Element> ¦ (<Action_Text>) ¦ ~<Action_Element>
       <Element> ::= <Operation> ¦ <Action> ¦ <Clause>
        <Clause> ::= <If> ¦ <Times> ¦ <While> ¦ <Until>
            <IF> ::= if (<Condition>) then (<Action_Text>) [ else (<Action_Text>) ]
         <Times> ::= times ( <Expression>, <Action_Text> )
         <While> ::= while ( <Condition>, <Action_Text> )
         <Until> ::= until ( <Condition>, <Action_Text> )
     <Operation> ::= <Operation_Name>( <Argument_List> )
        <Action> ::= <Action_Name>( <Argument_List> )
 <Argument_List> ::= <Argument> ¦ <Argument>, <Argument_List>
      <Argument> ::= "<Strings>" ¦ {<Expression>} ¦ <Frame> ¦ <Slot>
    <Expression> ::= <Algebraic_Expression> ¦ <Slot>
         <Frame> ::= #<Frame_ID> ¦ @<Frame_ID> ¦ $.<Frame_ID> ¦ $\\<Frame_ID> ¦ $
      <Frame_ID> ::= <Frame_Name> ¦ <Frame_ID>.<Frame_Name>
                       ¦ <Frame_Name>\\<Frame_Name> ¦ <Frame_ID>\\%%
          <Slot> ::= #<Slot_ID> ¦ @<Slot_ID> ¦ $.<Slot_ID> ¦ $\\<Slot_ID> ¦ $$
      <Slot_ID>  ::=  <Frame_ID>.<Slot_ID>
```

Fig.14 Syntax of Frame Operating Language

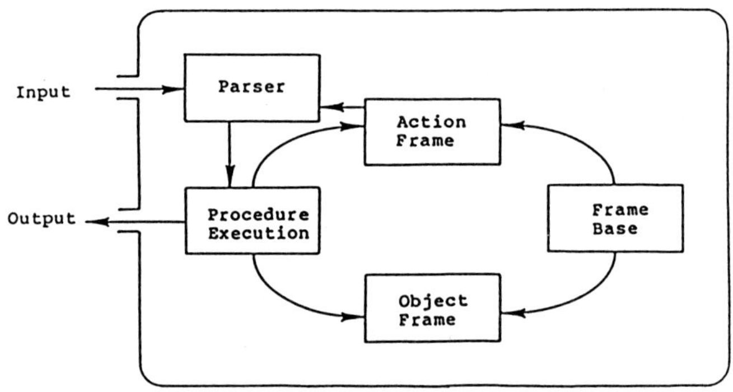

Fig.15 Frame Operations

4.2 Application of Frame System

Detail design is a process that translates the fundamental structure of a design object into the design solution which contains enough information for generating the corresponding entities. For this process, the following operations are required:

- determination of attributes of partial entities which compose fundamental structures.
- determination of relations between these partial entities.

Accordingly, such a system seems to be qualified for rationalizing this stage of design. In the upper stages, given specifications are translated into the network of function elements; partial entities corresponding to all function elements are assumed; relations which must exist between these partial entities are appointed (Figure 16). Since these stages can be automated by production rule systems, the problem of how to get the initial frames (which is the general problem of frame theory) is avoided in the detail design system.

There are two types of knowledge that are necessary for design: the static one and the dynamic one. Static knowledge is passive and is manipulated as referred to in design. Dynamic knowledge includes operations on design objects and their controls and also includes relations between attributes of design objects.

Machines can generally be represented as networks. In order to represent these structures, two types of frames are provided: Entity-frames correspond to nodes and Connection-frames to branches. By permitting Entity-frames to have sub-frames that describe lists of Connection-frames (that stand for connections between lower level Entities), it is possible to represent hierarchical network structures of machines (Figure 17).

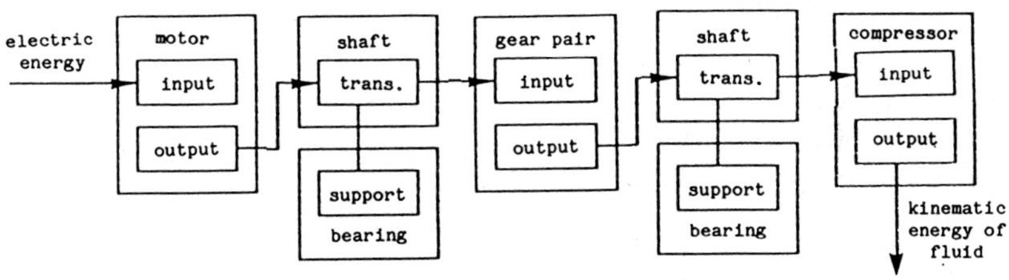

Fig.16 Fundamental Structure of a Machine

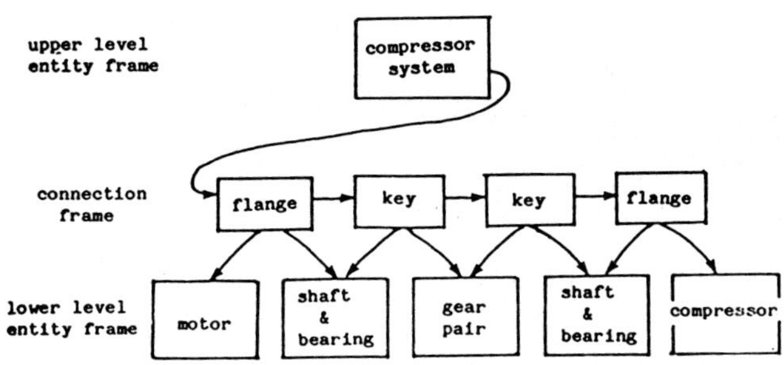

Fig.17 Representation of Machine Structure

In the actual stage of detail design, referencing of tables, such as standards, are given much weight. To represent these tabular data, sub-frames that have lists of the item values can be used.

On the other hand, dynamic knowledge means active knowledge that is concerned with the handling of design objects: it is described in the form of a procedure. It is possible to describe all procedures in Action Text, however, that approach is inefficient, because knowledge would become a mixture of knowledge about facts and processes that are empirically acquired. Accordingly, it is more adequate to add objective knowledge, that adheres closely to its objects, to the corresponding slots in the form of Attachments.

Attachments are classified into three types of knowledge: relations that must be satisfied by slot values, procedures that are necessary to fill the slot value and knowledge that doubles the parts of these procedures and relations.

Figure 18 shows an example of an Entity-frame which represents a screw including some Attachments. If some terms are set, a screw can be specified by proper procedures. Figure 19 shows one of those procedures described in an Action-frame, and the designing of a screw by evoking this Action-frame is shown in Figure 20.

```
name screw
class base
kind entity
nominal M10 DEFAULT
pitch 1.500000 DEFAULT
height 0.812000 DEFAULT
tanikei 8.376001 DEFAULT
material SF45
a_tens_str 0.000000 DEFAULT DEMON
 1       settbl(#materials,$.material)&curtbl($$,#materials)
tens_str 0.000000 DEFAULT DEMON
 1       $.pull/(pi/4*squ($.tanikei))
pull 2000.000000
```

Fig.18 Entity-frame "screw" and its Attachments

```
@@ fill($1.a_tens_str)&
@@ initbl($2)&
@@ curtbl($1.tanikei,$2)&
@@ fill($1.tens_str)&
@@ until(
@@       gtre({$1.a_tens_str},{$1.tens_str}),
@@       nextbl($1.tanikei,$2)&
@@       fill($1.tens_str)
@@ )&
@@ settbl($2,$1.nominal)&
@@ curtbl($1.pitch,$2)
@@ curtbl($1.height,$2)
```

Fig.19 Action Text in Action-frame "serfix"

```
>> frp(#scr1)
name scr1
class instance
kind screw
nominal M10 DEFAULT
pitch 1.500000 DEFAULT
height 0.812000 DEFAULT
tanikei 8.376001 DEFAULT
material SF45
a_tens_str 0.000000 DEFAULT DEMON
 1       settbl(#materials,$.material)&curtbl($$,#materials)
tens_str 0.000000 DEFAULT DEMON
 1       $.pull/(pi/4*squ($.tanikei))
pull 2000.000000

 YES!
ok
>> scrfix(#scr1,#metrscr)

 YES!
ok
>> frp(#scr1)
name scr1
class instance
kind screw
nominal M24
pitch 3.000000
height 1.624000
tanikei 20.751999
material SF45
a_tens_str 6.000000 DEFAULT DEMON
 1       settbl(#materials,$.material)&curtbl($$,#materials)
tens_str 5.91368 DEMON
 1       $.pull/(pi/4*squ($.tanikei))
pull 2000.000000
```

Fig.20 Execution of Action-frame "scrfix"

REFERENCES

[1] Yoshikawa, H., General Design Theory and a CAD System, in: Sata, T., Warman, E. (eds.), Man-Machine Communication in CAD/CAM (North-Holland, Amsterdam, 1981).

[2] They were published in SEIMITSUKIKAI (Proc. of Japan Soc. of Precision Engineering) in Japanese. Yoshikawa, H., Mechanical Topology, 38 (1972) 1018; Introduction to General Design Theory, 45 (1979) 906; Rules of Structure in Machine, 45 (1979) 1099.

[3] Yoshikawa, H., Multi-purpose Modeling of Mechanical Systems, in: Bjørke, Franksen (eds.), System Structures in Engineering (Tapir, Trondheim, 1978).

[4] Yoshikawa, H., CAD Framework Guided by General Design Theory, in: Bø, K., Lillehagen, F.M. (eds.), CAD Systems Framework (North-Holland, Amsterdam, 1983).

[5] Tomiyama, T., Yoshikawa, H., Requirements and Principles for Intelligent CAD Systems, in: Gero, J.S. (ed.), Knowledge Engineering in Computer-Aided Design (North-Holland, Amsterdam, 1985).

[6] Takeshige, A., Tomiyama, T., Yoshikawa, H., An Application of Frame System to CAD, in: Hubka, V., et al. (eds.), Proc. of International Conference on Engineering Design (Heurista, Zuerich, 1985).

IMPACTS

ARTIFICIAL INTELLIGENCE IN MANUFACTURING
T. Bernold (editor)
Elsevier Science Publishers B.V. (North-Holland)

CHALLENGES AND FORSEEABLE CHANGES IN THE INDUSTRIAL MANAGEMENT SYSTEM

E.C. STOTKO

Manager
CIM Programme Office IBM Europe
Munich, F.R.G.

Introduction

Management is the skill, art and science of getting things done.

In an industrial environment, the most common measurement of management performance is the ROI achievement.

As industry evolves and market dynamics change, the management system has to be adjusted to cope with new sets of conditions and new political, social, economic and technological environments.

In previous phases of the industrial evolution, management attention was focused successively on achieving stable production operations, organising the production flow to best utilise labor (Taylorisation), introducing new materials and manufacturing processes, mechanising production to grow output volumes, increasing product complexity.

All these developments happened in days of a steady pace of production and long product life cycles.

In Europe this period of production (output) orientation ended in the 1960's when the replenishment demands after World-War II were satisfied.

The change in the industrial management system began, when companies were forced to re-orient themselves towards the market and more volatile public tastes.

New technologies enabled new company initiatives. Information Technology has become the most pervasive of them, since it shows up in new products, in factory- and office processes, in the communication network, which connects company resources and in new methods and tools for management and professionals which enable an enterprise to behave like a SYSTEM. A much greater effectiveness of the entire business operation will be the consequence, if and when the new technologies become properly embedded in the non-technical environments of people and organisations.

Whenever industry enters a new era like now that of the COMPUTER INTEGRATED MANUFACTURING ENTERPRISE, significant adjustments have to be made.

The management team is particularly challenged to negociate the necessary turns and adjustments successfully. FORESIGHT and vision or NATURAL INTELLIGENCE is asked for, before tools of Artificial Intelligence can be employed gainfully.

The necessary changes are not primarily of a technical nature. Rather a CULTURAL CHANGE in the business management system is required.

For instance, the traditional management hierarchies of passing orders down and reporting back up the line, will have to give way to hierarchies of delegating responsibilities and of free two-way communication across the enterprise as necessary to accomplish a task effectively.

Improvement goals should no longer only focus on PRODUCTIVITY in a narrow quantitative sense, but rather more on the quality or EXCELLENCE of ideas, plans, professionalism and results.

Management is challenged to establish new orientations, new qualities of behaviour, new goal views and associated measurements.

Especially in regard to future applications of experts systems and AI tools it must be understood that, irrespective of their inherent capabilities, they will never produce useful answers if they are not asked the right questions.

This paper was intended to set the stage for the subsequent papers, which deal with specific aspects of and experiences with AI techniques and tools.

Normally, a method or tool has little or no value by itself; rather its value is derived from its proper use by people within an organisation.

Especially in the industrial environment today, the offer of "wares" abound, which all claim significant improvements to present operations. In their sum total, benefit claims exceed by far the 100 % of the possible.

So, not loosing sight of the forest for all its trees is of vital importance for proper management decisions.

The Industrial Scene

Strong competition on a world-wide basis, dynamic market behaviours, cost and inflexibility of the workforce and growing government interference characterise the types of pressures european industry is currently facing. Countermeasures lie in product innovation, creative marketing, customisation, better planning, cost reduction and more effective overall organisations. New types of co-operation and efforts towards diversification are evident.

More than in the US or Japan, Europe faces also a conflict of interests between INDUSTRY and the actions it must take to maintain or regain competitiveness, the INDIVIDUAL and its expectations from society regarding employment, self-ful-filment and quality of life and ENVIRONMENT protectionism.

The Implications of the CIM Enterprise (CIME) or the Information Technology (IT) Era of Industry

Like in the previous phases of industry, one key technology dominates other secondary ingredients of change.

In the case of I/T it pervades all facets of industrial operation.

1. It has become an indispensible element of PRODUCT ENGINEERING. Electronic capabilities add product functions, flexibility and ease of use, and overall appeal.

2. MANUFACTURING PROCESSES are driven, monitored and controlled through micro processors, digital sensors and real-time computer communication and control systems.

3. OPERATIONAL COMPANY PROCESSES like Planning, Engineering, Administration and Decision Support are significantly improved through computer aids.

4. The means of real-time, digital communication improve OVERALL BUSINESS EFFECTIVENESS, by bringing back transparency of company operations and allowing faster response to changes. The global impact of CIME can and should be measured in its effect on the three key dimensions LEAD TIME, QUALITY and COST.

   These effects become achievable through technology, but only if it is embedded in a receptive non-computer environment of people, organisational structure and management schemes.

   Company processes are dynamic in nature. They require co-operation of people and equipment across departmental borders to produce meaningful results.

   Information systems can establish temporary communication links among the resources involved, supply them with the required data and collect the information content of the work results.

5. Finally, Computer-based Communication networks allow closer and stronger links between a manufacturing company and its suppliers and distributors.

A dynamic state network with interactive graphic modeling and simulation capability is a highly promising technique for the purposes of points 2 to 5, with AI implications.

But the initial improvements of operational efficiency come from the implementation of information systems with the following three common sense type capabilities across the whole company "wall-to-wall":

1. Communicate operational data in digital form from one area to another and from source to destination.

2. Generate significant INFORMATION from raw DATA in a way, that the momentary needs of a given situation are met.

3. Control operational data against pre-determined objectives (plans).

The essence of a future CIM Enterprise is, that the real-life physical ACTION WORLD of the company is given its matching abstract INFORMATION WORLD as a mirror image equivalent, in an organised, consistent fashion.

That way consequences of contemplated plans can be analysed, simulated and evaluated as abstract information models at little expense and short time delay. Only once a plan has then been validated and fine-tuned, will the physical action world be charged with the transformation of plans into products, processes, and services.

Challenges of Industrial Management

The challenges of industrial management are manyfold. Key items which have to be addressed are: rate of product and process innovation, product quality, manufacturing flexibility and product customisation, product cost, overall organisational response and effectiveness.

But no singular measure will produce dramatic effects. Management must conceive of a multitude of complementary actions and drive them forward in concert along a strategic master plan.

The measures are reflected in a set of operational assumption for a future CIM-Enterprise:

1. The CIM-Enterprise will function as a SYSTEM. The roles and tasks of the individual resource, notably people and machines, become more clearly defined, hence more planable and predictable.

2. communication and routine control become basic system functions. Feed-backs permit more automatic situation assessments.

3. The scientific base of operations will grow in scope and depth. Large scale re-qualification programs become necessary, to upgrade the knowledge level of employees.

4. The qualification profile of people will change from:

   quantitative to qualitative tasks
   execution to supervision
   physical strain to mental stress.

5. The time gained through use of routine support tools, should be used to improve plans and decisions. Not displacing people is an aim of a CIM system, rather improving results through the opportunity to perform more design alternatives and deeper penetration of each design. The goal is excellence over productivity.

6. Knowledge bases will permit systematic "organisational learning" from doing, thus making better use of the immaterial asset "know-how" and leading to a sort of "dynamic standardisation" i. e. an evolution of recommended best practices based on earned experiences.

7. Such a CIM System not only relieves management from tedious tasks, it eliminates the need for many layers of middle

management alltogether. This helps regain transparency over what is going on and timely corrective actions when necessary.

8. Two layers of management however, will gain importance:

   On one side the think-tank manager with the ability to lead a company in the right strategic direction. On the other side the first line people manager. He is responsible for providing action space and creating a work environment and climate which motivates people to creative and quality work.

9. The wide range of CA-X tools will allow the individual person to accomplish a larger work content. This will lead to "De-Specialisation" and job enrichment.

   This "direct" team will take care of the necessary "indirect" work itself. This is part of the delegation of responsibility for complete tasks and results. Less need for potentially harmful management interference exists.

10. The CIM environment of numeric digitalisation will automatically improve quality of results. Self-checking becomes a feature of the system.

11. Investment decisions have to be prepared more carefully, in light of higher spending levels and growing risks of wrong decisions.

12. The management orientation will change

FROM	TO
Direct supervision of people	Giving people tasks to accomplish by themselves
Hierarchical communication	Multi-mode communication in Q-circle fashion giving people selfidentity and pride of accomplishment

Team leadership capabilities become most important. Many present management tasks will be assumed by information systems, including routine decision making.

Future Management Perspective

Presently, management is still largely ad hoc reaction. Human problems, priority conflicts, meetings, telephone calls, appointments, urgent memo's to top management are typical of a manager's world. Diagnosis, short term and long term implications are processed into individual decisions and actions.

In future the following changes in the managment system will occur:

1. Strategic company goals and directions are clearly established.

2. The new co-operative management culture will be established and "lived", which means parting with some cherished old priviledges.

3. The planning efforts are strengthened to achieve a best possible consistent set of plans which permit achieving the established goals, in place of previous cut-and-try practices.

4. Organisations will be stream-lined along "business processes".

5. ENGINEERING-MANUFACTURING-SALES-SERVICE will be conceived as one common SYSTEM, within which each function is a SUB-SYSTEM.

   In addition, the CIM system extends beyond company boundaries to include suppliers and distributors.

   On-going work on an open CIM-System architecture (CIM-OSA) will pave the way towards improved co-operation.

6. Measurements and decision criteria must be developed, which allows dynamic tuning of each sub-system operation to achieve best combined results. This means overcoming selfish departmental attitudes and reconciling conflicting interests of various functions in an "automatic" way by interconnecting the individual control loops.

7. After best possible plans are established, the organisation and its resources must be "energised" to execute the plans efficiently.

8. For best individual performances, a work environment must be created, which gives people action space and motivation.

9. Encourage a "less art, more science" approach to tasks and provide the necessary CA-x tools as well as the pre-requisite training, and establish effective communication networks.

10. Manage the transition from present to future system.

   In so doing, provide foreward looking leadership and give employees a strongly positive perspective of the future.

In this brief scenario of a future management system there is room for a generic AI application.

It is that of a real-time knowledge-based monitor of the on-going business processes, with the inherent ability of learning from doing.

<u>Action Space of Management</u>

Having discussed the new company set-up in the CIM age, let us now focus on the action space available to management.

As in the past, a company derives its base of existence from the successful balancing of interests of three distinct groups: The CUSTOMERS, SHARE HOLDERS and EMPLOYEES. Each group has its own motivations of supporting the company's well-being, but also its own expectation of sharing the fruits of success.

Beyond this, CIM opens up new avenues of competing for business. Especially small and medium-size companies stand to gain relative strength over larger enterprises.

- Economies of scope allow a company to produce an infinite

variety of products within the technological capabilities of flexible, data-driven manufacturing equipment.

- Possessing such capabilities allows to actively pursue customisations from product design to delivery.

- Product innovations can be brought to market faster because of the numeric data chains from CAD to CAM.

Pre-requisites to be established by management are:

- Strategic orientation

- Product family planning consistent over Engineering, Manufacturing process range, sales, service and finance.

- New accounting and cost management methods. Because of significantly changed cost structures and cause/effect relationships conventional accounting methods are no longer applicable.

- New measurements of performance to encourage the desired new attitudes and behaviours.

The fundamental three dimensions of performance remain:

TIME as needed for:

- new product development
- order throughput
- reaction to changes
- decisions

QUALITY as reflected in:

- error prevention
- error correction
- degree of optimality of plans and results

COST as occuring in:

- investments
- products
- material
- labor
    - direct
    - indirect

Proper trade-offs within and across these three dimensions constitute the art and more and more the science of successful management.

The Management-By-Horizon Concept

In addition to existing management schemes, I propose the new scheme of Management-By-Horizon (MBH).

It is based on the basic recognition, that plans and actions step through distinct STAGES with unique characteristics, time horizons, and impacts on business results.

Six stages are at present distinguished:

1. GOAL AND DIRECTION SETTING

2. PLANNING

3. ENGINEERING

4. SET-UP

5. EXECUTION/PRODUCTION

6. RESULTS MONITORING AND ANALYSIS

Stages 1 through 3 can be combined into FUTURES MANAGEMENT, stages 4 and 5 into PRESENCE MANAGEMENT.

The latter one is charged with "milking the cow", that is, make the most out of given investments into products, manufacturing equipment, skills, sales and service organisations. The presence management deals with physical processes of material transformation. It generates revenue from the sales of products and services.

FUTURES MANAGEMENT by contrast is charged with securing the future, i. e. development of new sets of products, processes and sales and service capabilities, ready to take the place of or complement the present sets, whenever they have lost market attractiveness.

An important aspect of MANAGEMENT-BY-HORIZON is that, as one zooms in on the presence through the stages, one cannot recover on a later stage from errors or omissions committed at a previous stage.

In other words, the action space and the possible impact on business success gets successively smaller as one moves from goal setting to execution. Another aspect is that of an ORBITAL WINDOW.

Each stage has a certain latitude of fault tolerance with freedom of action, within which actions can occur independently of other stages without jeopardising the stage results. This action space constitutes the frame for measurement of management performance.

The transition from one stage to the next is comparable to a quantum leap, it is discontinuous.

The continuity across the stages is provided by the common "mission", which is to be accomplished.

The Managment-by-Horizon concept may serve as a frame of reference for the implementation of CIM Systems for individual enterprises and complete sub-industries.

CLOSING REMARK

With all the discussion of new technologies and techniques it should not be forgotten that technologies and systems make things possible, but it is PEOPLE that make it happen.

People should not be driven by 'SYSTEMS', rather they should mastermind their proper and beneficial use.

ARTIFICIAL INTELLIGENCE IN MANUFACTURING
T. Bernold (editor)
Elsevier Science Publishers B.V. (North-Holland)

CAPACITY PLANNING AND SYSTEM DESIGN IN MANUFACTURING OPERATIONS

Professor Gabriel Bitran
Devanath Tirupati

Sloan School of Management
Massachusetts Institute Technology
Cambridge, MA 02143 U.S.A.

Dr. Themis Papageorge

Palladian Software, Inc.
Four Cambridge Center
Cambridge, MA 02143 U.S.A.

Manufacturing operations have always been challenging because of the complexity and interrelationships of the manufacturing process. Traditional information systems collect historical data effectively but have not been successful in providing assistance in planning and design. The appropriate advanced operations management techniques require experts to apply and interpret the results. Expert systems technology can be applied to derive functional relationships among capacity, work-in-process and lead times. Our results show that artificial intelligence applied to manufacturing may be the key to manage the complexities associated with the decision-making process in that field.

INTRODUCTION

The management of manufacturing operations has been an almost intractable problem since the industrial revolution. The reasons for that are both external, and inherent. Some of the external reasons are:

- Multiplicity of products and operations, forcing tradeoff decisions among capacity, technology, work-in-process and lead times.
- Competition, continuously altering the demand pattern of a manufacturer and thus making long-term forecasting a complicated endeavor.
- Technological change, accelerating the rate of machine obsolescence and changing the rate of labor productivity.
- Evolution of knowledge, enforcing constant adjustment and reevaluation of operating practices.
- Multiple operating environments, such as batch, continuous and assembly, often coexisting in the same plant.
- Multi-plant and multinational considerations, such as sourcing and distribution, subjecting operating decisions to spatial, time and inter- national accounting constraints.
- Computer network issues, impacting both intra- and interplant operating decisions. Furthermore, manufacturing/operations problems are also inherently complex.
- Modeling by using deterministic techniques, such as mathematical programming, leads to NP-Hard type of problems.

- Modeling using simulation techniques, such as Monte Carlo, leads to combinatorially explosive situations.

The approach adopted by most researchers and practitioners has been the hierarchical decomposition of the total problem in related but independent subproblems. In our paper we concentrate on the subproblem of capacity planning within a single plant, dealing with the external reasons of demand changes, technological changes, changes in product mix and part policies and with the inherent problems of combinatorial explosion. We examine the design of an expert system for capacity planning in manufacturing, the operation of the system, and some initial results from the application of the system.

SYSTEM DESIGN

The state-of-the-art in system design for manufacturing has changed rapidly in recent years, primarily due to the advancement in computer hardware and software and the recognition of fertile application areas. The great majority of systems, however, are transactional in nature and collect historical data. The MRP II type of systems have been in operation for a number of years with mixed results. Although extensions of these systems have been used as planning systems, inherent design limitations and the complexity of the problem itself rendered little success to such efforts.

Figure 1 presents a schematic design for today's manufacturing systems. Data Collection systems and order entry systems feed the Shop Floor Control database, the MRP (if it exists) system and the Scheduling system which in turn drive the plant. Although this system design has been applied extensively, it has some important deficiencies:

- Operational data are subject to error, whereas transactional systems assume 100 percent accuracy for their operation.

- Transactional systems do not provide for uncertainty in demand and other critical factors.

- Capacity constraints are not explicitly considered, and lead times are assumed to be constant; any change in these parameters leads to time consuming computer runs.

Artificial intelligence technology, and specifically expert system technology, present an opportunity to overcome complexity and provide the means to accurate and timely planning. Expert systems incorporate heuristics and experiential rules enabling the system to identify functional relationships among important planning variables and enhance the utility of historical data.

Figure 2 presents a schematic system design for manufacturing incorporating expert systems technology. In this case, the expert systems component is responsible for medium- and long-term planning, frequently exchanging information with the transactional systems which are responsible for short-term operations. This new system design has some important advantages:

- "Semi-accurate" transcational data are acceptable for planning, because the expert system uses heuristics and rules at a higher level of aggregation.

- Functional relationships among capacity, lead times, work-in-process inventory and utilization are established by the expert system and are constantly fed back to transactional systems without a need for time consuming and expensive trial-and-error computer runs.

- Historical data are aggregated in product families and machine centers (Group Technologies and FMS), processed through a capacity planning module and scheduled at an aggregate level.

- Capacity planning becomes the central issue of manufacturing planning at the medium- and long-term levels. First, because the system helps users identify and manage physical assets such as capacity. Second, assuming effective product families/machine centers analysis and capacity planning, the problem of scheduling becomes easier to address.

A. Theoretical Background

Our approach to the problem of capacity planning has been to use the basic concepts of the theory of networks of queues. We have extended the procedures in the literature to more general networks. This extension is based on expert rules adjusting for arbitrary distributions and multiple products, which provide superior approximations to real life manufacturing planning situations.

Furthermore, we have introduced explicit targeting procedures which utilize a set of greedy heuristics and provide the capability of solving the network for the final and intermediate values of a targeted variable, and all variables functionally related to it, without actually performing a combinatorially explosive number of simulations.

The network representation of the plant that we use is a function of product families and machine stations. Machine stations are objects comprised of one or more machines of which one is critical and may cause a bottleneck. Product families are objects comprised of one or more products sharing similar routing and/or processing and setup times. This object oriented formulation allows for hierarchical decomposition of a company into plants, and then plants within plants at any appropriate level of detail. The arbitrary degree of abstraction in our system design allows for multiple levels of manufacturing planning.

The computational efficiency of our formulation, employing machine centers, product families, networks of queues with expert rules and greedy heuristics is such that our calculations are faster by many orders of magnitude as compared with event simulators. Furthermore, the complexity of our computations increases only linearly with the number of machine stations and product families, whereas the complexity increases exponentially with an event simulator. Finally, our computational experiments indicate that the accuracy of our calculations is in the 90 percent range for 90 percent of the stations, and the total error for the system is in the 0.5-5 percent range.

B. Measurement Issues

The measurement of the properties of the machine stations and product family objects is critical to every manufacturing system. Although our expert system design can handle more errors and uncertainty than a transactional system, measurement is an unresolved issue. For purposes of experimentation we used a real life manufacturing database with good results. The following should be considered as issues for further research and development:

- Capacity increments are usually measured in machine or labor hours, regular and overtime. For large increments of capacity, additional shifts or new machines may be required. The accounting of these changes is neither simple nor accurately done in most organizations, especially when multi-year investments such as new machines are considered.

- Technological change and experience effects impact the measurement of capacity and labor costs in non-trivial ways.

- Work-in-Process measurements at each machine station are dependent on the value added at previous stations. Specific valuation rules need to be developed.

- The non-linear relationship between utilization, work-in-process, lead times, and even quality make the measurement of utilization critical to manufacturing planning. Therefore, variable loads on the floor and changing capacity availability affect utilization measurements.

SYSTEM OPERATION: AN APPLICATION

The expert system design we have developed emulates the expertise of a human planner/advisor in a complex manufacturing environment. The managers of a manufacturing operation are expected to make decisions which affect the entire organization. For example:

- Capacity utilization affects the ROI of a company and consequently its stock price, cost of capital and ability to expand

- Lead times affect the response to the market of a company and consequently its revenue stream, market share and ability to compete

- Work-in-Process Inventories affect the working capital of a company and consequently its cash outflows, cost structure and profitability.

In order to allow the user to address some of these questions meaningfully, the system encourages the user to group products into product families and machines into machine centers. In our application we have successfully reduced a semiconductor manufacturing operation with 900 products and 200 machines to a network of 11 product families and 13 machine stations. Figure 3 shows the graphic representation of some of these objects on a computer screen; the figure shows, for a given product family, the stations visited and the sequence in which the product families flow through the shop. These features of the system reduce the complex semiconductor facility to a workable abstraction level, exactly as a human planner/advisor would do.

The next step is to analyze the network. The system provides a natural way for the user to describe the demand patterns of the product families, processing sequence, corresponding characteristics and the intended type of analysis. The system is capable of discerning arbitrary probability distributions; the analysis includes heuristics, approximate formulae and sets of expert rules and an event simulator for situations where greater accuracy is required. Figure 4 shows network performance statistics such as WIP in units and dollars, lead time in weeks, yield in percent and utilization in percent for each machine station and for the total system for a specific product family.

This preliminary or base case analysis serves as a starting point for the user. The manager or the planner of this semiconductor manufacturing facility can deal with complex issues such as:

- The impact of setup times on WIP.
- The optimal flow of materials and semifinished goods or how a layout change may affect the production flow.
- Lowering machine utilization and increasing labor productivity may reduce WIP and production costs leading to higher ROI.
- Allocate labor to more than one process.
- Reduce WIP and lead times.
- Establish a leveled production schedule on the basis of the functional relationships between product families and machine centers, capacity and lead times.

The capacity planning methodology that we have developed when applied to the specific semiconductor manufacturing facility, shown in Figures 3 and 4, led to a number of important conclusions:

- Product families and machine stations were rationalized.
- Plants within plants were discovered.
- The production flow was rationalized and resulted in minor engineering changes which simplified greatly the manufacturing process.
- Lead times were reduced by 10 percent.
- New capacity addition scenaria were analyzed quickly and accurately, without having to perform numerous, time consuming simulations.
- New technologies and product mix changes were analyzed and prioritized.

CONCLUSION

A new methodology for capacity planning and system design in manufacturing operations using expert systems technology has been developed and applied to a specific manufacturing facility. This methodology compliments existing manufacturing information systems by offering heuristic approaches to reducing complexity and evaluating alternatives.

Functional relationships among capacity, work-in-process and lead times have been established in general and can be applied to a variety of manufacturing operations at several levels of hierarchical decomposition. Numerical results show that this methodology can combine accuracy, speed and human-like advisor characteristics, such as reasoning from inexact data and at various levels of abstraction.

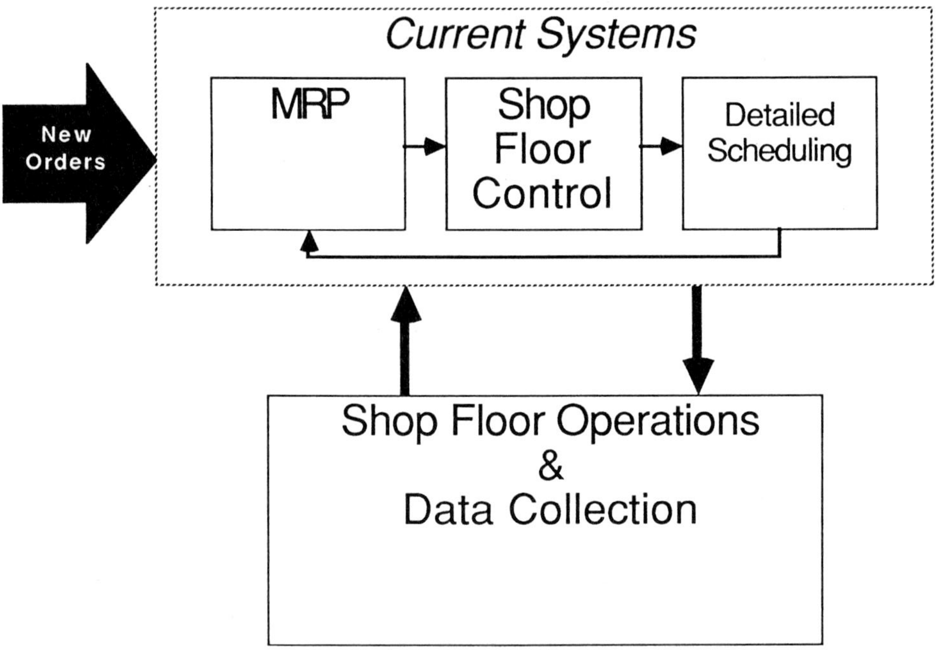

FIGURE 1

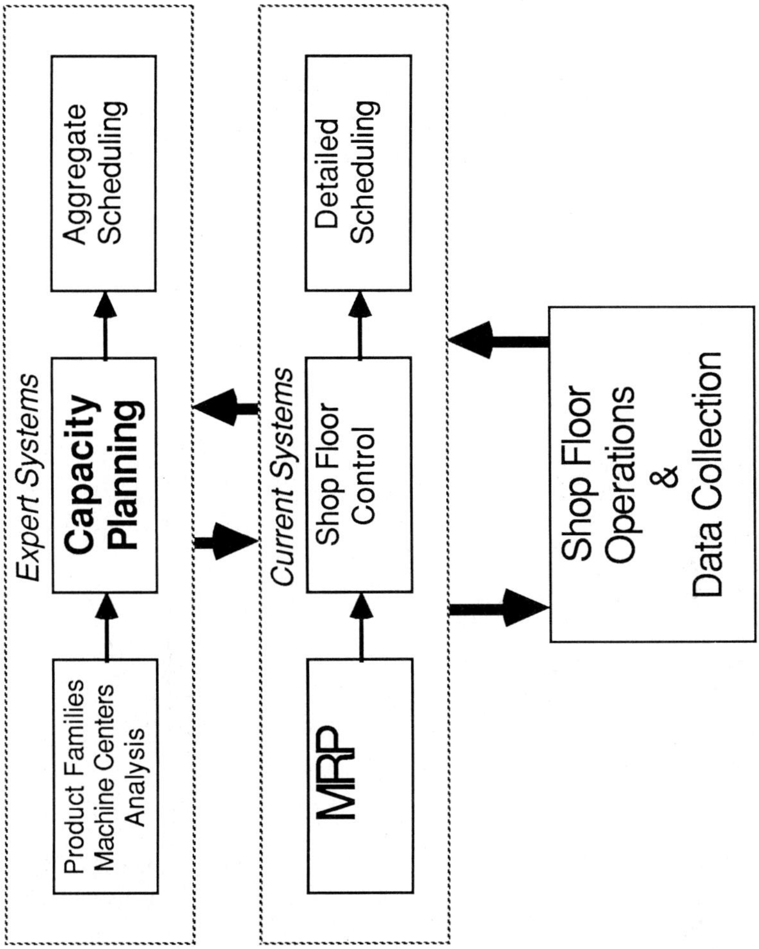

Expert Systems
Product Families
Machine Centers
Analysis
Capacity
Planning
Aggregate
Scheduling
Current Systems
MRP
Shop Floor
Control
Detailed
Scheduling
Shop Floor
Operations
&
Data Collection

FIGURE 2

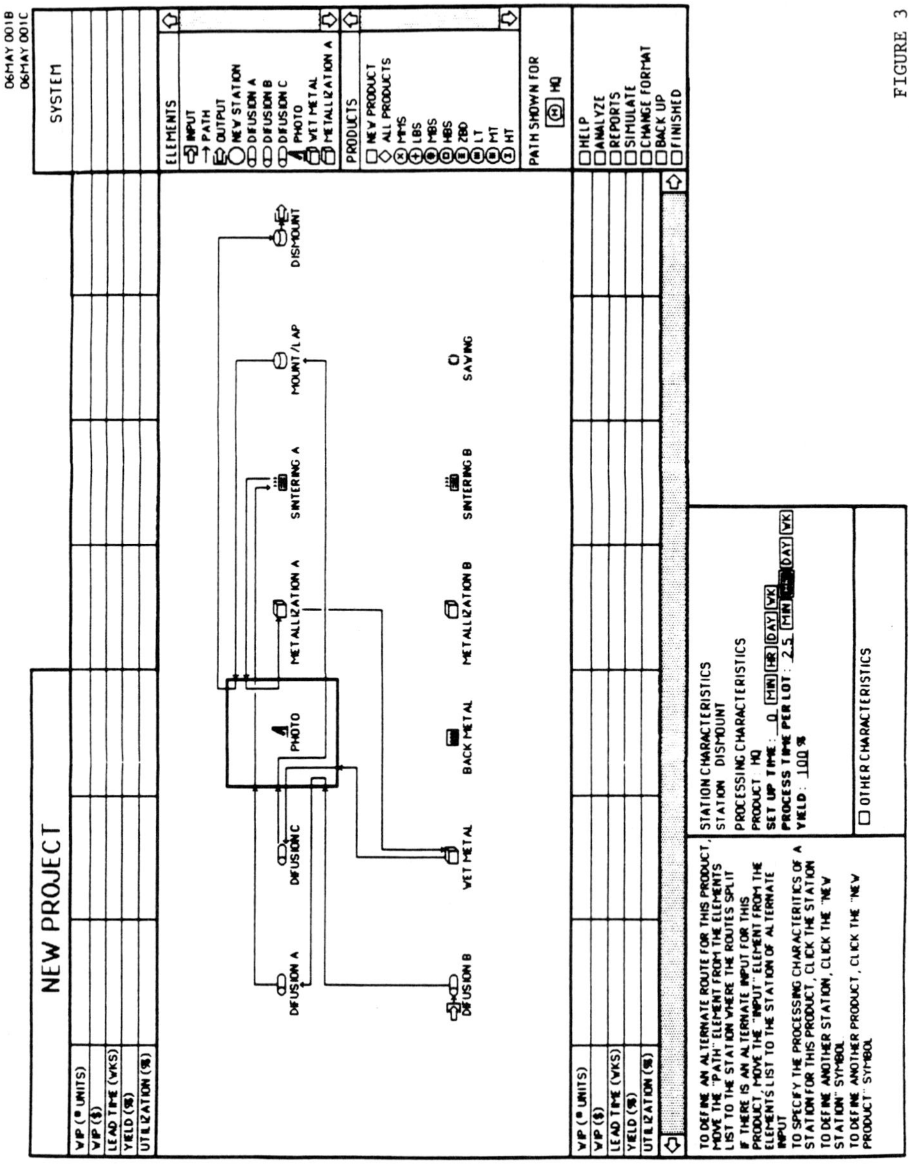
06MAY 001B
06MAY 001C
NEW PROJECT
SYSTEM
WIP (# UNITS)
WIP ($)
LEAD TIME (WKS)
YIELD (%)
UTILIZATION (%)
ELEMENTS
INPUT
PATH
OUTPUT
NEW STATION
DIFUSION A
DIFUSION B
DIFUSION C
PHOTO
WET METAL
METALLIZATION A
PRODUCTS
NEW PRODUCT
ALL PRODUCTS
MMS
LBS
MBS
HBS
ZBD
LT
MT
HT
PATH SHOWN FOR
HQ
HELP
ANALYZE
REPORTS
SIMULATE
CHANGE FORMAT
BACK UP
FINISHED
DIFUSION A
DIFUSION C
PHOTO
METALLIZATION A
SINTERING A
MOUNT/LAP
DISMOUNT
DIFUSION B
WET METAL
BACK METAL
METALLIZATION B
SINTERING B
SAWING
TO DEFINE AN ALTERNATE ROUTE FOR THIS PRODUCT, MOVE THE "PATH" ELEMENT FROM THE ELEMENTS LIST TO THE STATION WHERE THE ROUTES SPLIT
IF THERE IS AN ALTERNATE INPUT FOR THIS PRODUCT, MOVE THE "INPUT" ELEMENT FROM THE ELEMENTS LIST TO THE STATION OF ALTERNATE INPUT
TO SPECIFY THE PROCESSING CHARACTERITICS OF A STATION FOR THIS PRODUCT, CLICK THE STATION
TO DEFINE ANOTHER STATION, CLICK THE "NEW STATION" SYMBOL
TO DEFINE ANOTHER PRODUCT, CLICK THE "NEW PRODUCT" SYMBOL
STATION CHARACTERISTICS
STATION DISMOUNT
PROCESSING CHARACTERISTICS
PRODUCT HQ
SET UP TIME: 0 MIN HR DAY WK
PROCESS TIME PER LOT: 2.5 MIN HR DAY WK
YIELD: 100%
OTHER CHARACTERISTICS

FIGURE 3

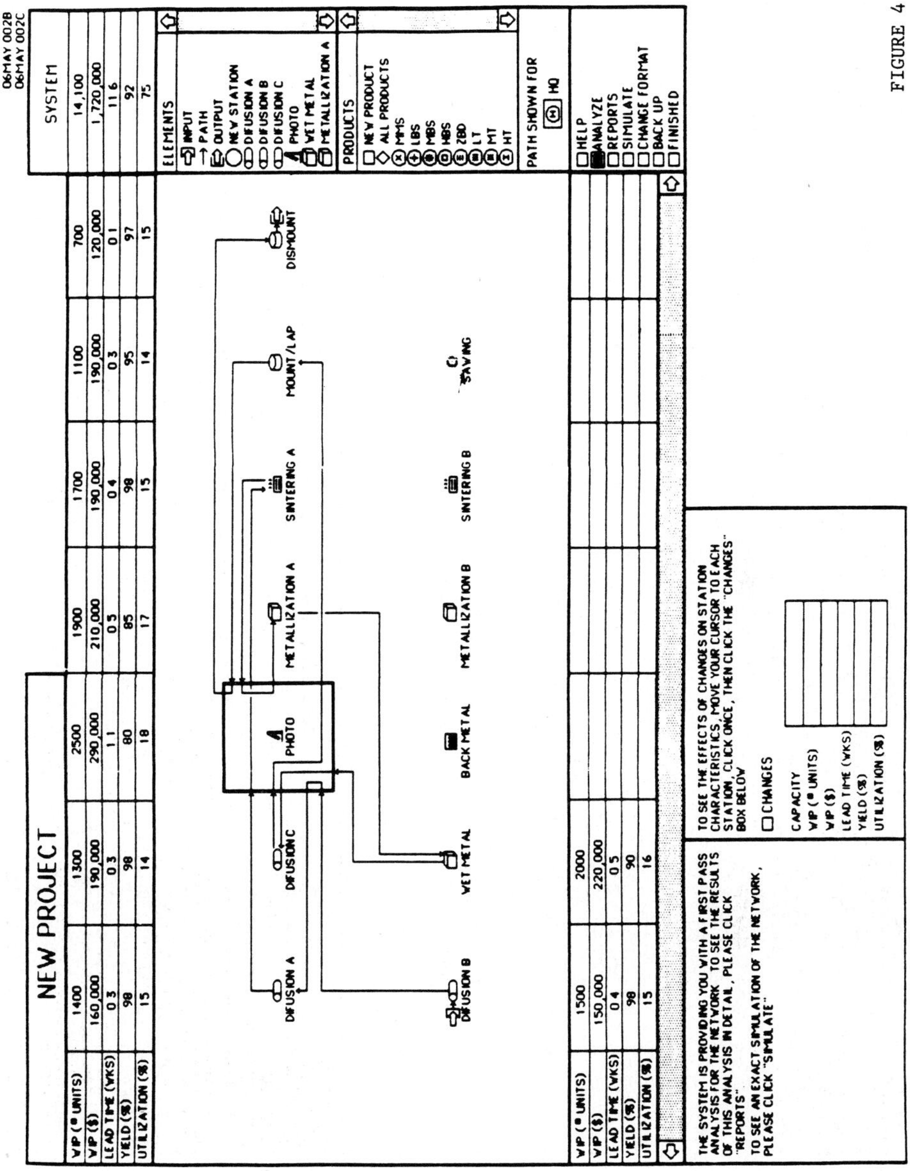
NEW PROJECT
SYSTEM
06MAY 002B
06MAY 002C
WIP (# UNITS)
WIP ($)
LEAD TIME (WKS)
YIELD (%)
UTILIZATION (%)
DIFUSION A
DIFUSION B
DIFUSION C
PHOTO
METALLIZATION A
METALLIZATION B
SINTERING A
SINTERING B
MOUNT/LAP
DISMOUNT
WET METAL
BACK METAL
SAVING
ELEMENTS
INPUT
PATH
OUTPUT
NEW STATION
PRODUCTS
NEW PRODUCT
ALL PRODUCTS
PATH SHOWN FOR
HELP
ANALYZE
REPORTS
SIMULATE
CHANGE FORMAT
BACK UP
FINISHED
THE SYSTEM IS PROVIDING YOU WITH A FIRST PASS ANALYSIS FOR THE NETWORK. TO SEE THE RESULTS OF THIS ANALYSIS IN DETAIL, PLEASE CLICK "REPORTS".
TO SEE AN EXACT SIMULATION OF THE NETWORK, PLEASE CLICK "SIMULATE".
TO SEE THE EFFECTS OF CHANGES ON STATION CHARACTERISTICS, MOVE YOUR CURSOR TO EACH STATION, CLICK ONCE, THEN CLICK THE "CHANGES" BOX BELOW
CHANGES
CAPACITY

FIGURE 4

ARTIFICIAL INTELLIGENCE IN MANUFACTURING
T. Bernold (editor)
Elsevier Science Publishers B.V. (North-Holland)

SYSTEMS OF THE 5TH GENERATION AND THEIR IMPACT ON THE QUALIFICATION, TRAINING, AND RETRAINING OF MANAGEMENT

Hans-Jochen Schneider

Institut für Angewandte Informatik
Technische Universität Berlin
1000 Berlin 10
Fed. Rep. of Germany

The introduction of 5th generation systems means that new tasks must be assumed by management. Information will become a product. The industrial society is on the way to an information society. Information has to be evaluated and must be introduced in balance and profit and loss sheets. Information managers must be members of the board of directors. The concept of an information center allows the concurrent handling of decentralized organization and data processing structures and central information gathering. Expert systems will support managers mainly in planning and decision making. Modelling and simulation of company behavior will be utilized as an important means for decision support. Management has to be introduced to the capabilities and limits of flexible and knowledge-based systems through continuous education.

INTRODUCTION

The installation of new knowledge-based manufacturing systems - i.e. systems of the 5th generation - will lead to a significant increase in industrial productivity and flexibility through late customizing of the products. The development in industrial production in most application areas will result in Computer Integrated Manufacturing (CIM).

What does a typical CIM-solution look like? First, a new part is defined geometrically with an interactive graphics program for Computer Aided Design (CAD) and Manufacturing (CAM) running on a host system. The corresponding data needed for production planning will be collected in the bill-of-material system (BOM). The data will be manipulated by a system for order entry, material planning, and control running, also on the host computer system.

Robots controlled by robot language expressions and the geometric and bill-of-material data downloaded from the host system to a cell-controller and robot controller then do (for instance) the mechanical assembly or the component insertion. Additional information about joining points of the various manufacturing components (e.g. transfer stations, transport system, bowl feeders, magazines) and sensor data support this process. After the order is completed with administrative data, all manufacturing data (numbers of good and faulty parts, consumption of subparts, execution time needed) are sent back from the cell-controller to the production planning system in order to initiate delivery notes and invoices printing. The cell-controller could typically be a personal computer. It is used to create, compile, and download the robot language expressions to the robot controller. Fig. 1 shows the data flow and components in such a CIM-system.

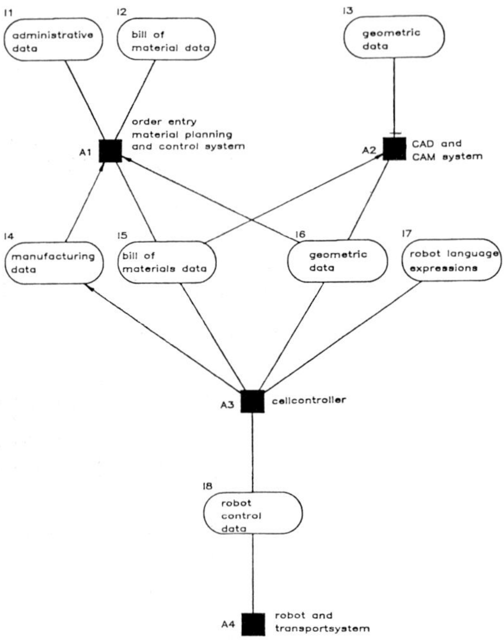

Fig. 1: Data flow in a CIM-system

The Abraham (ISAC) Description Technique

Symbol	Interpretation	Denotation
■	Activity (agency) – action – process – information processing – data processing – program execution – event structure	A_i (P_i)
⬭	Information unit (channel) material unit – message representating component – carrier of messages – carrier of material – communication medium – state structure	I_i (M_i)
│ ↑	Access paths describe the interrelationships between the activities and the information units	
┆□┆	Compound activity describes an activity of the (n – 1) th level on the n – th level (n ≧ 2)	
⬭ (dashed)	Compound information unit describes an information unit of the (n – 1) th level on the n – th level (n ≧ 2)	

Knowledge-based manufacturing systems of the 5th generation will allow an increase of profit and revenue in the face of further sharpening competition and rapidly changing technologies. These systems have to provide a link between the four major islands of automation: engineering design, production planning, manufacturing control and factory automation including material handling. Fig. 2 shows a functional and dataflow diagram of the CIM-process. Artificial Intelligence concepts and expert systems technology will support the planning, installation, and running of such complex systems.

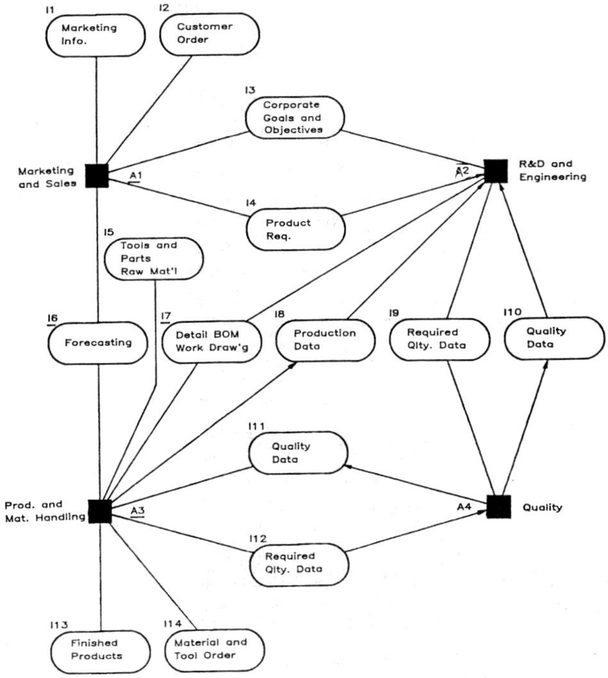

Fig. 2a: Functional diagram of the manufacturing enterprise

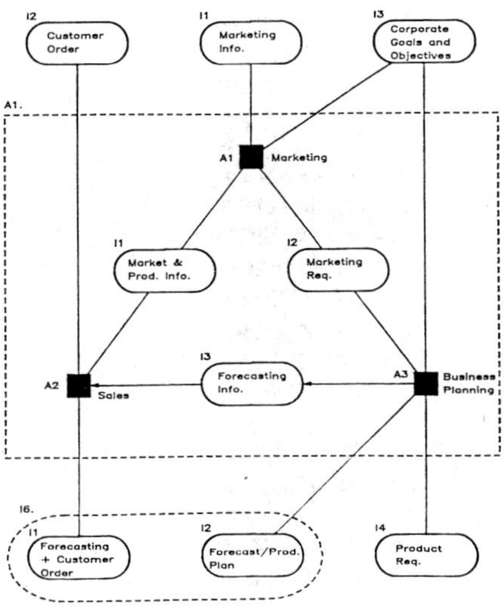

Fig. 2b: Refinement of agency "A1: Marketing and Sales"

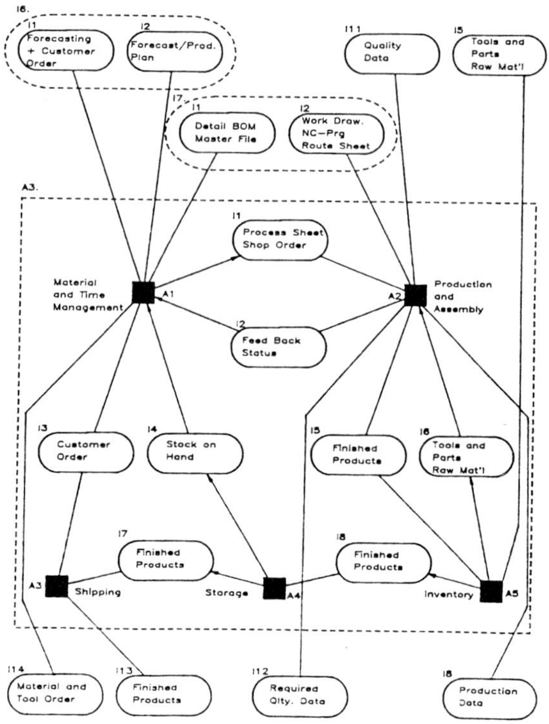

Fig. 2c: Refinement of agency "A3: Production and Materials Handling"

Up to now CIM has been meant as a strategy. A CIM-product is not available, yet. There are several practical implementations and theories of how to introduce a CIM-strategy in a company. The main approaches known are Manufacturing Resource Planning (MRP II), KANBAN of the Toyota Motor Company, and just-in-time production. Further approaches have to be developed because of the disadvantages of the existing ones.

Before describing CIM-systems of the 3rd and 4th generation, I would like to present several theses about the data processing market:

> Thesis 1: More and more companies utilize data processing internally and in their products. Therefore data processing is a matter of top management.
>
> Thesis 2: The industrial society is in the transition to an information society. Information must consciously be viewed as a product. Information has to be evaluated and consequently introduced in balance and profit and loss sheets.

Thesis 3: The speed of acting on the market will be the crucial measure of competence of a company in the future. The fastness of information gathering, processing, and transfer is an unconditional prerequisite for the secured existence of a company.

Thesis 4: All functional units of a company will be part of an integrated company information system. Decentralized organization and data processing will be coupled with centralized information management.

Thesis 5: Software, not hardware, has become the decisive cost and evaluation factor when introducing and following-up the data processing solutions.

CIM-SYSTEMS OF THE 3RD AND 4TH GENERATION

One of the main observations of the past five years in data processing is the integration of the areas commercial data processing, text processing, engineering data processing (CAD, CAM, CAE: Computer Aided Engineering), process control, and telecommunication. As a consequence the various information islands in the company are linked together, yielding computer integrated manufacturing (see Fig. 3). All information units concerning order entry, product design, analysis and simulation, process control, quality assurance, material handling, assembly inspection, test, inventory control, and shipping are handled by the same database.

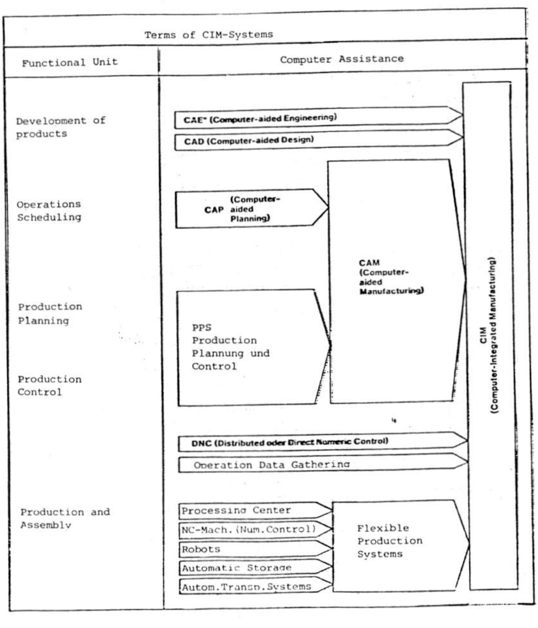

Fig. 3: Information islands are linked together

The various types of input and output are growing together into a multi-functional-workstation, a depiction which is shown in Fig.4.

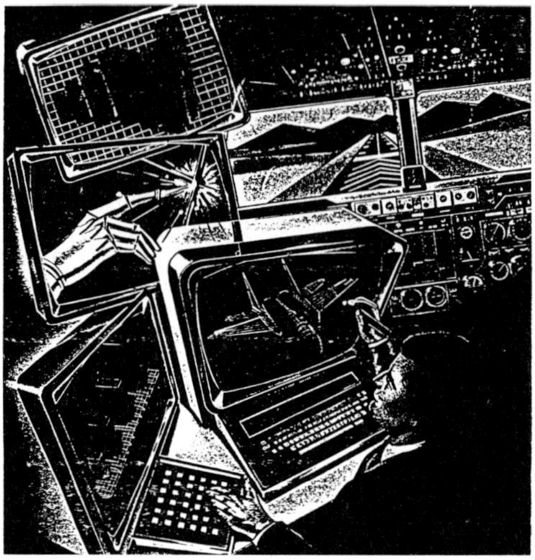

Fig. 4: A multi-functional workstation

The integration of the whole information flow in a company implies that more and more departments and organizational units must be involved in running a project and finding a problem solution. Fig. 5 shows the situation in an example project.

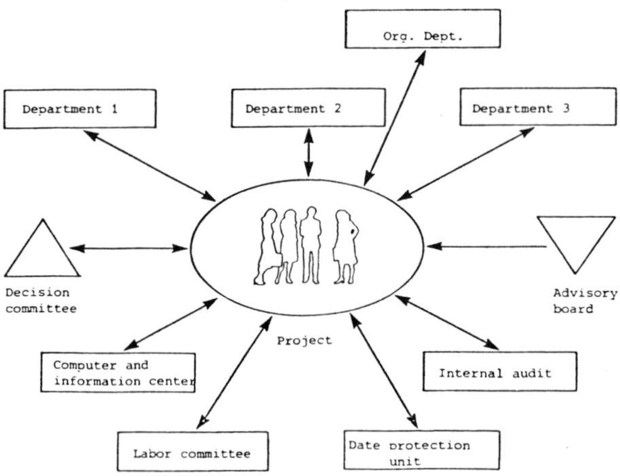

Fig. 5: Involvement of different types of people in a project

Computer integrated manufacturing requires management to have more overview on the one hand and more detailed knowledge on the other hand in order to be able to evaluate given problem situations and to make correct decisions. The tasks of management are becoming more and more like the tasks of architects. Top managers must be the company architects of the future.

KNOWLEDGE-BASED SYSTEMS OF THE 5TH GENERATION

Up to now single technologies such as micro-electronics, software technology, fiber-electronics, and telecommunication have been in the center of gravity. In the future, computer technology, integrating methods, tools and strategies to develop systems will be emphasized.

The Japanese Program for Fifth Generation Computer Systems (FGCS) has received a fully unexpected worldwide echo. It is the first time that the combined government and industry research and development activity of one country has triggered so many follow-up research activities in other countries. With this program Japan aims to be number one in the world in the information technology market in the beginning and middle of the 1990's. The MITI (Ministry of International Trade and Industry) has established a ten year program of research and development in the field of information technology. It also initiated a new institution related to complex research, the ICOT (Institution for new generation COmputer Technology). More than 450 researchers will be fully employed in 1986 in this program.

Other countries reacted to this program:

- The DARPA program on Strategic Computing and Survivability was submitted to the US Congress in November 1983.

- The US National Science Foundation announced a new initiative for the development of super computers in November 1983.

- The US Department of Defense announced a long range plan for systems of the "Nth Generation" in 1983.

- Fifteen well-known American companies have initiated a new company: Microelectronics and Computer Technology (MCC). This company will have a yearly budget of 50-100 million US dollars and is meant as an answer to the Japanese challenge.

- The European Commission has started the ESPRIT Program (European Strategic Program for Research and development in Information Technologies). The first five year plan covers a budget of 3-4 billion DM for research and development.

- The British government has started the ALVEY program which is very similar to the ESPRIT program.

One of the main goals of the Japanese FGCS programm is to supplement classical data processing by knowledge processing. This goal requires a deep insight in human processes like problem description, problem understanding, problem solving, and very closely related subject areas like knowledge representation, knowledge processing, knowledge acquisition, dialog handling, and the analysis and synthesis of natural language. New formal calculi, artificial languages, and methods have to be developed to solve all these problems. In the past few years special logical theories, new logical programming languages (PROLOG), knowledge representation and processing languages, and system architectures have been developed.

The goals of the FGCS-program are very high. If they can be reached a series of milestones will be overcome:

- The application specialist will develop his own application programs.

- The maintenance of manually specified and (semi-) automatically generated software products will be much easier because of the unique programming style and the standardized interfaces.

- The problem solution quality (the highest goal available) will be improved drastically because the users can concentrate only on the solution process and will not be irritated by data pro-cessing details.

- Software development will be an engineering discipline, (the classical art of tailored software development will become obsolete).

- Systems will be able to recognize failures, inconsistencies, and discrepancies/contradictions and in most cases they will themselves be able to correct and resolve the mistakes automati-cally.

- Systems will contain powerful tools for knowledge acquisition (i.e. knowledge transfer from the expert to the system), that are fast, easy, and reliable.

- Systems will offer a substantial increase in programming productivity.

The Japanese researchers see the following subject fields as the main application areas of Fifth Generation Computer Systems:

- Automatic translation systems of natural language texts with a vocabulary of more than 100,000 words and an accuracy of translations of about 90%;

- expert systems in various aplication areas with 5,000 different words and 10,000 inference rules;

- knowledge-based software development environments;

- knowledge-based VLSI (Very Large Scale Integration) chip production and CAD (Computer-Aided Design), CAM (Computer-Aided Manufacturing), CAE (Computer-Aided Engineering), CAP (Computer-Aided Planning) systems, and robot systems;

- knowledge-based problem solution systems for specific application areas. E.g. in office automation: automatic dictating machine with a vocabulary of 10,000 different words and speaker identification out of a sample of 500 different speakers. In addition there will be important applications in medical diagnosis, in juridical and tax query systems, in chemical structure analysis, in detecting new oil and mineral sources, in failure detection within systems.

In Fig. 6 the original architecture of the fifth generation computer systems developed by Moto-Oka and his research colleagues is represented. The hardware architecture of these computer systems is not more von Neumann architecture. In the software field they are also going in unconventional directions.

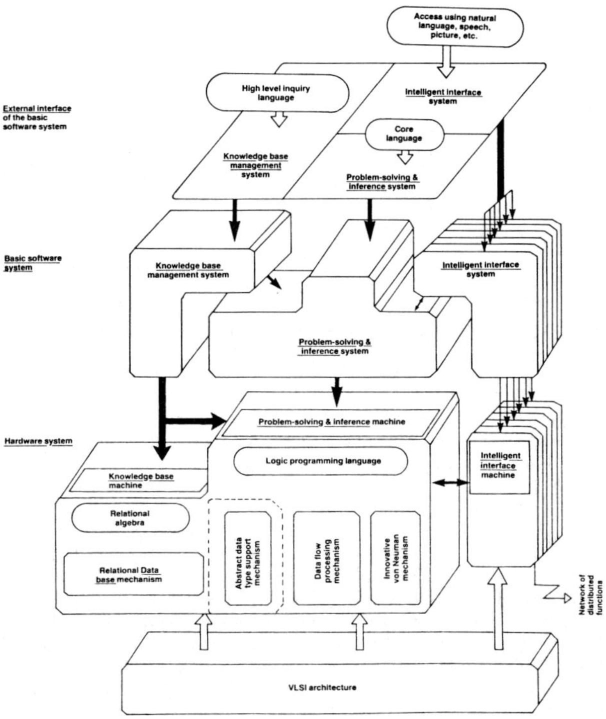

Fig. 6: Architecture of Fifth Generation Computer Systems (FGCS)

The requirements of FGCS for hardware speed lie between 10^{12} and 10^{14} instructions per second. The database machines should be capable to store 100 to 1,000 billion bytes. The intelligent interface language, 2,000 grammar rules, and about 10,000 different graphic elements.
The main components of this architecture are represented in Fig.7. This picture shows the most important software research and development areas:

- problem solving
- knowledge-based processing
- intelligent interfaces.

Hardware specialists will build the first layer in VLSI technology; software specialists will develop the corresponding software components and interfaces for the casual user.

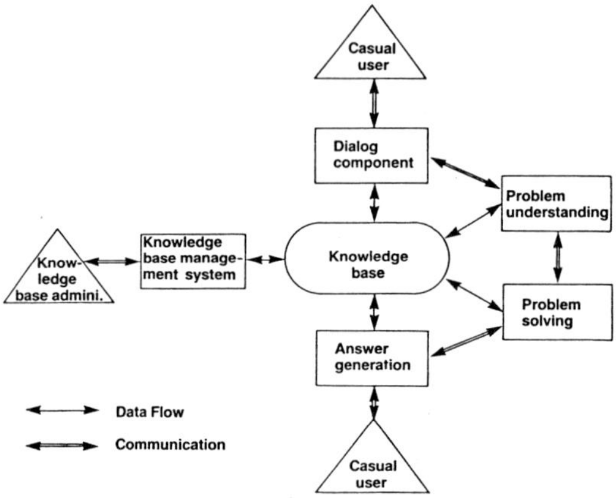

Fig. 7: Dataflow diagram of a knowledge-based system

Fig. 7 shows a dataflow diagram of a knowledge-based system for general problem solving. The dialog component needs concepts and techniques from linguistics, logic, artificial intelligence, and cognitive science to be able to map natural and quasi-natural language expressions into formal representation-language expressions and vice versa. The casual user will have a very user-oriented natural interface.

The problem-understanding component analyses the content of the language expressions. It must represent and process spatial and temporal knowledge, it must differentiate between facts, events, and actions, and it must construct a model of the real world, its objects, and dialog partners. This component must recognize whether a given string is e.g. a proposition, a query, or an instruction. Corresponding to this recognition process, a command procedure for database input and/or change, query, or program generation must be created/generated. It must evaluate the transferred knowledge, whether it is vague, reliable, evident or relevant knowledge. The component must recognize failures and contradictions and must protect confidential and secret information.

The problem solution component then processes e.g. the database transactions, the translation to lower levels of abstraction, the program generation, and the movement of the robot arms. Then the component delivers the result in internal form to the answer generation component.

The answer generation component transforms the internal results into a quasi-natural language form and interacts in a user-friendly manner with the casual user.

The knowledgebase administrator is responsible for the very important knowledgebase; he maintains, creates, and changes the structure and the content of the knowledgebase.

Fig. 8 shows the components of an expert system for a specific subfield of manufacturing.

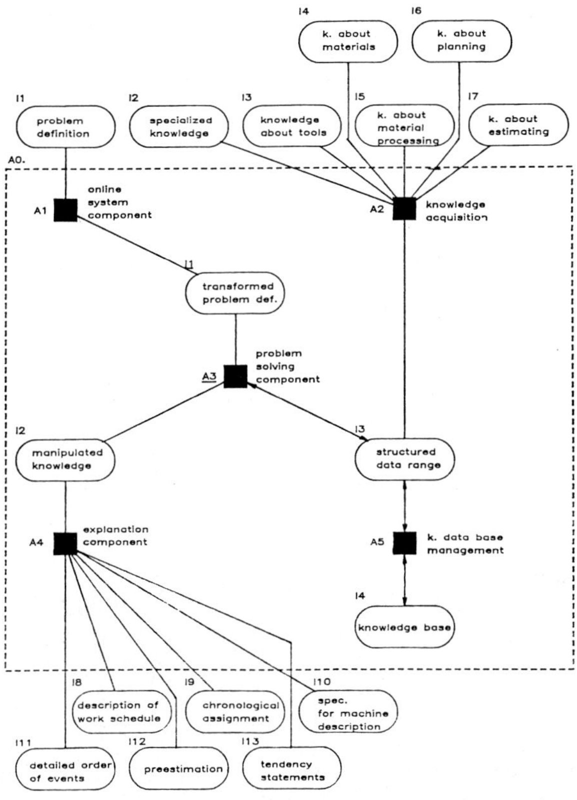

Fig. 8: Functional units of an expert system

CONCLUSION: THE INFORMATION NEED OF MANAGEMENT

It is essential to create the position of an information manager in companies. He should be responsible for the whole information flow within the company and for introducting concepts and tools of information technology in the products of the company. He must be part of top management and head of the information center in which all information about the company, its products, the market, the existing and future technologies, and legislation flow together. Furthermore he must coordinate the cooperation of the various departments, consultants, organizational people, works council, revisers, and those responsible for data protection.

The available information allows an exact modelling of the company on the strategic and tactical levels and the simulation of alternative decision strategies. Existing methods and software tools from econometrics, operations research and system dynamics must be extended and developed.

REFERENCES:

/BO82/ Bonner, S.; Shin, K.G.: Comparative Study of Robot Languages. COMPUTER 5(12), 1982

/DI84/ Diebold Management Report No.8/1984

/HA80/ Habel, Ch.; Rollinger, C.-R.; Schmidt, A.; Schneider,H.-J.: A Logic-oriented Approach to Automatic Text-Understanding in: Bolc, L. (ed.): Natural Language Based Computer Systems, Hanser-Verlag, 57-117, 1980

/MO82/ Moto-Oka, T. (ed.): Fifth Generation Computer Systems, North-Holland Publishing Company, 287 pages, 1982

/RO83/ Rollinger, C.-R.; Schneider, H.-J.: Text Understanding as a Knowledge-based Process, in: IRFIS-Proceedings,Heidelberg, 1983

/SCN79/ Schneider, H.-J. (ed.): Formal Models and Practical Tools for Information Systems Design, Proc. of the IFIP-WG 8.1, Oxford 1979, North-Holland, 306 pages, 1979

/SCN82A/ Schneider, H.-J.; Wasserman, A.I. (eds.) : Automated Tools for Information Systems Design, Proc. of the IFIP-WG 8.1 Working Conference, New Orleans, January 1982, North-Holland, 268 pages, 1982

/SCN82B/ Schneider, H.-J.: Techniques and Formal Tools for Design, Realization and Evaluation of Evolutionary Information Systems. In: Hawgood, J.: Evolutionary Information Systems, Proc. of the IFIP-TC8 Working Conference, Budapest, September 1981, North-Holland, 13-37, 1982

/SCN82C/ Schneider, H.-J. (ed.): Distributed Databases, Proc. of the Second International Symposium on Distributed Databases, Berlin, September 1982, North-Holland, 375 pages, 1982

/SCN84A/ Schneider, H.-J.: An Integrated Concept for Information System Development, Proc. CAE Symposium at Jiatong University Shanghai, 1984

/SCN84B/ Schneider, H.-J.: Software der fünften Generation - die Lösung der Softwarekrise? - In: Wissenschaftsmagazin der TU Berlin zum Thema "Kommunikation", Heft 6, 80-84, 1984

ARTIFICIAL INTELLIGENCE IN MANUFACTURING
T. Bernold (editor)
Elsevier Science Publishers B.V. (North-Holland)

5TH GENERATION SYSTEMS AND NEW TRENDS IN MANAGEMENT

Jörgen Biedermann

Comor AG, Computer- und Organisationsberatung
Seefeldstrasse 62
CH - 8008 Zürich

PANELDISCUSSION

During the panel discussion on the implementation of artificial intelligence (AI) in manufacturing, changes in management structures as well as the abolition of management levels were foreseen by many different participants. There were, however, basic reservations concerning the implementation of AI, in that one should not presuppose absolute knowledge and allround capability. The view was put forward that it was wrong to start with the assumption that organisational problems could be solved with the help of AI, as the roots of such problems lay primarily in the area of human relations. The crucial emotional tasks of management could not, under any circumstances, be dealt with by means of AI.

In response to the question: what management needs to know about AI in general and in detail, it was pointed-out the increasing complexity of production and product manufacturing could only be supervised with the aid of AI. This gave rise to the question as to whether Man's own inability to recreate the world around us ought to be replaced by machines. One of the limitations of AI is the absence of emotion, which can be seen as supportive of isolationary tendencies in the carrying out of management tasks. On the other hand, it was questionable whether in the running of a modern enterprise the type of manager with brains and sensibility will still be required or if the technique of AI will be given too high a priority.

Developers and users of flexible manufacturing systems will promote the introduction of AI for total system servicing, as the present single-operator manufacturing method will become very quickly over-extended. In this respect, it is not the organisation that must be changed to accomodate AI but rather the AI system should be introduced as an extra resource so that free space will be left at the operational level for the operator to use his humanistic, social capabilities to the full. This requires, as opposed to today's current practice, new production systems which can foresee the corresponding interaction levels which effectively support decision-making at the place of work. In general, the question of how will AI change our management structures is turned around so that possibilities for modifications to management structures will be decided with the help of AI. As a starting point, hierarchical structures will be shown as dynamic and not static concepts, which also emphasises the fact that existing management structures are always defeated by the criticism of people in the organisation itself.

AI may be defined as a medium which not only stores large amounts of unintelligible information but which can also make it immediately usable, so that management must learn how to use it. Moreover, the hope must be expressed that these systems themselves develop a level of creativity, or at least can support it. In contrast to this, it will be maintained that the borderline between man and machine is not really known, and that reality in general cannot be formalised, but only a model of that reality. As an example of the difference between man's capability and that of the AI machine one may cite the chess programme as a simple two-person, zero-result game with comprehensive information. The superiority of man over machine is in his selective capability and associative powers, but man does not yet know how these decision processes function. Thus, the employment of complex problem solving tasks where there is incomplete information, as occurs in most real situations, has yet to be carried out. It is recommended that a start should be made where AI can give reliable information and not to seize hold of most complex problems right away. In the complementary situation between man and auxiliary means, AI can mutually influence creativity as the highly developed stystem requires an ever-increasing level of creative input, which also leads to a further development of the system. These development processes, however, require time and so current expert systems should not be over-valued as they are just at the beginning of their development cycle. On the other hand, anticipation of later possible solutions can lead to a loss of transparency for management and so to a devaluation of AI, as has happened in the case of Operations Research and Manufacturing Information Systems.

Analogous to earlier development, the division of work between man and machine may be carried out by mechanical means. Equally, man's intellectual capabilities can be divided up by the AI machine. This leads to the controversial question whether it will be good for mankind that such a technique, which lies in a purely mechanical area, should be used for the historical division of labour, or whether it should be suspended. One must accept, however, that these questions should be asked. The danger of automation with AI will be in the loss of capability and knowledge, in the sense of 'know-how', that a practice being carried out by automation is no longer necessary. Here it is recommended that the experience and knowledge of a healthy, human understanding is brought into AI science, or that a corresponding free-space is maintained.

The resulting demands on managerial skills are directed towards stronger attitudes to the future: flexible, knowledge-based and throughout, technically orientated. To illustrate, one could draw comparisons with architects and orchestra conductors where the effectiveness of performance results from cooperation. Here, the strength of information, knowledge and understanding are of paramount importance. The AI stystems which are yet to be developed will support management. With the transition of various phases of development to a system of automation, an overlapping form of coordination is recommended so as to avoid any loss of 'know-how' or experience, and conversely to achieve a synergy effect. Such an AI system should then promote and strengthen human participation without replacing or mechanising it. An essential criterion for such a development is the question of whether the private sector enterprises can provide secure jobs and actively participate in them. A great number of practical applications will be necessary in order to show that AI can become an active part of management and not suffer the same fate as operations research which now belongs to a branch of theoretical science.

ARTIFICIAL INTELLIGENCE IN MANUFACTURING
T. Bernold (editor)
Elsevier Science Publishers B.V. (North-Holland)

FORECASTS FOR THE USE OF ARTIFICIAL INTELLIGENCE IN THE USA

Michael T. Knasel

Automated Systems Division
The MK-Ferguson Company
Cleveland, Ohio 44114
U.S.A.

The use of artificial intelligence in manufacturing has finally emerged as a reality in the United States. Buoyed by the general growth of trained computer scientists, the lower cost hardware and software designed for easier user input, a number of experimental programs are under way to exploit artificial intelligence for manufacturing purposes.

It is a pleasure to address you today on the uses of AI in manufacturing. My talk will cover forecasts of the use of AI in the United States. The data I have compiled is based on a several year investigation of the applicability of Artificial Intelligence (AI) in manufacturing, and other industrial applications. The framework of my efforts were as a user of AI in a factory situation, rather than in an office setting.

The initial questions that I asked were directed toward uncovering practical uses of AI that would stand the test of time. I also sought to define the product forms, and determine the delivery systems and distribution channels. AI has had a quarter of a century to mature to the point where it can be tried for industrial use and I was eager to know if it was going to be a success or not.

Based on a two-year study of enduser needs, vendor research and development thrusts, and product capabilities, we have compiled a comprehensive picture of the emergence of artificial intelligence as a legitimate market in the United States. Our findings are discussed below.

While little of the current year AI expenditures of about $ 330 million in the U.S.A. is directed at manufacturing specifically, we believe that by 1989, up to forty percent (40%) of AI activity will be in exploring applications in manufacturing. If, as our research indicates, AI is developed as a real time control system for processes and factory-wide control purposes, then AI installed and in use in manufacturing could grow significantly in the 1990 to 2000 decade.

Overall I found that there would be three ingredients for success of AI applications. First, management must have the committment and staying power to introduce the new technology and to see it mature. Secondly, the AI products have to serve a need in terms of their performance and price. Finally, the enduser staff must have the background and training available to assimilate AI technology into their enterprise.

Our study covered user activities; their successes and failures. We analyzed vendor products and tried out techniques on site. Analog reasoning from other industrial ventures was used to create scenarios as to the future of AI development from todays base of expertise.

AI has evolved beyond the pioneer innovators of the 1960's time frame as manufacturing firms such as Digital Equipment began to apply the techniques. Many others became "window shoppers" to see if AI was useful. This has spawned token AI computers and AI researchers on the corporate staffs of many major enterprises. Better computers, and especially better software in the expert system area has opened up a good many initial opportunities to try out AI, and industry is experimenting with a wide range of applications. Exhibit 1 illustrates known categories of present day exploration. Note that several areas such as design, planning, monitoring, process control and the like are high value industrial jobs. The essential feature of AI is to use heuristics to augment a current method that may be inadequate due to the intractability of the data, or the inability to create a physical model of the process.

EXHIBIT 1

Expert System Functional Application Areas

- Design of systems, equipment, etc.
- Planning/scheduling
- Monitoring of equipment
- Testing
- Inspection
- Diagnosis
- Repair and maintenance
- Instruction
- Image interpretation
- Interpretation of data
- Advice
- Information retrieval
- Process control
- Evaluation
- Development of new knowledge from experience

The first major conclusion of our efforts was that:

Manufacturing endusers have the widest spectrum of AI applications.

While financial, office and natural resources and closely related tasks will benefit from AI, the manufacturing floor offers the greatest challenge. This is because of the high value added associated with process improvement. AI for real time control is an excellent, yet largely untapped application. Eventually tying AI to robots, machine vision, and voice recognition will create unusual capability in manufacturing.

There are numerous examples of AI in early development and test in factories. Exhibit 2 lists nine examples out of 20 to 30 publically announced activities. Generally these are all expert systems; the area of AI in the widest field use in manufacturing.

Beyond todays use, we could forecast an excellent future for AI use, due to the potentially high value added, when applied sucessfully.

EXHIBIT 2

U.S. Companies and Institutions Active in Manufacturing Applications of AI

CAM-I	-	3d solids modeling generative planning
CINCINNATI MILACRON	-	expert technology process plan
CAMERON IRON WORKS	-	group technology process plan
CARNEGIE GROUP	-	process planning
DEERE CORPORATION	-	feature based sheetmetal process plan
HUGHES AIRCRAFT	-	electronic assembly work process design
GENERAL MOTORS	-	minority interests in AI and machine vision firm
LOCKHEED GEORGIA	-	plan generation
USAF	-	research and development institute

The growth of AI development in industry is still too fragile, and it will not flourish unless there is a "champion". There appears to be such a catalytic effect due to U.S. Federal Government interest in AI, our second finding relates to the belief that:

U.S. Government programs will stimulate continued development and commercial spin-off.

The U.S. Government is accelerating its committment to, and confidence in, AI techniques. This goes beyond the use projected in military systems, and includes ambitious plans for the use of AI in space programs. AI is planned for manufacturing applications by the U.S. Air Force and they hope to form an AI in Manufacturing Research Institute shortly. The inescapable conclusion is that the manufacturing area is a ripe one for growth.

Our third area of investigation sought answers to our question: "How are AI products developed and delivered"?

AI delivery systems are now involved and require a protracted time to build and learn.

Experts or "masters" are thinly stretched over the enduser base. Exhibit 3 illustrates the process we have identified, a three phase development, fueled at the top by government and large "high tech" corporations such as we have today in aerospace and automotive markets. The central focus on universities to train "masters" was a bottleneck that is being broken (in phase 2) as the companies shift resources to new start-up ventures. At the same time certain public domain commercial products (such as versions of common LISP) spin-off.

In phase 3, packaged tools are available for developers in factory settings to work with experimental data finally creating an applications system that can be used by a company with ease.

EXHIBIT 3

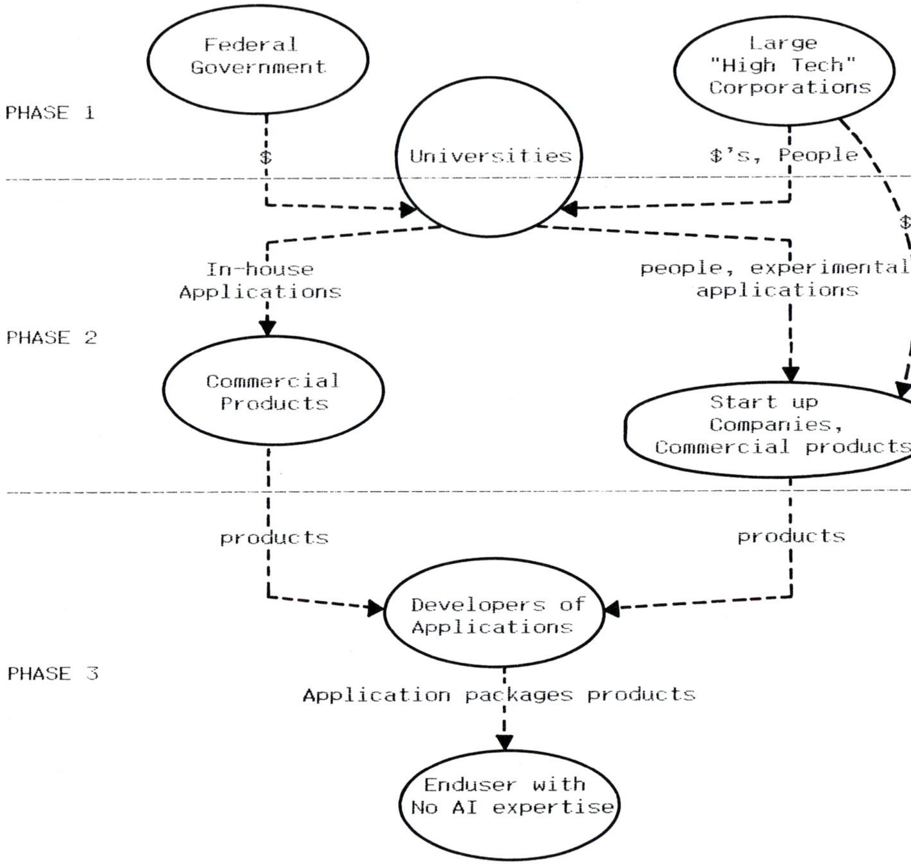

The fourth major conclusion was then, that:

<u>The growth in AI after 1990 will depend on development of applications that do not require user programming input.</u>

The enduser community capable and willing to program AI software will plateau, and growth will be possible only for AI applications that provide value for the price. Applications in office automation (outlining letters, reports, and creating data bases), financial, auditing and legal areas appear promising.

A thorough review of the "players" in the AI marketplace in the United States reveals that nearly 50 firms have announced entry. The most prominent ones are listed in Exhibit 4 along with a suggested infrastructure of new start-ups from software distribution, and larger companies for which AI is as of now only a sideliner, albeit an increasingly important one. The computer industry maxim "software sells hardware" has already caused most of the factory computer market leaders to office AI products, usually through third party AI specialty houses.

EXHIBIT 4

AI Industry Structure - U.S.A.

NEW COMPANIES

- Artificial Intelligence Corporation
- Carnegie Group, Inc.
- Cognitive Systems, Inc.
- Computer * Thought Corporation
- ExperTelligence, Inc.
- Inference Corporation
- IntelliCorp
- KDS
- LISP Machine, Inc.
- Logicware, Inc.
- Software A&E, Inc.
- Smart Systems Technology
- Symbolics, Inc.
- Teknowledge, Inc.

SOFTWARE DISTRIBUTORS

- Expert Systems International
- Frantz, Inc.
- Gold Hill Computers, Inc.
- Jeffrey Perrone & Associates, Inc.
- Production System Technologies, Inc.
- Programming Logic Systems, Inc.

LARGE COMPANIES

- Apollo Computer, Inc.
- Data General Corporation
- Digital Equipment Corporation
- General Research Corporation
- IBM
- Microsoft
- PERQ Systems Corporation
- Radian
- Sun Microsystems, Inc.
- Systems Research Laboratories, Inc.
- Tektronix
- Texas Instruments, Inc.
- Xerox Corporation

At the beginning of the study, the common belief in the U.S. was that AI would be a 5 to 20 billion dollar market by 1990. But we found that analysis did not support this bullish estimate, because it neglected important issues such as:

- volume pricing,
- market saturation, and
- internal competition among various AI techniques.

We define the AI market solely as AI computers (i.e. computers with special symbolic processors, or sold only to support AI applications), AI software and AI services. Engineering workstations for example, are not included. Although they are used to execute AI software, they are purchased for broader uses than solely AI. We exclude robots, machine vision and voice recognition since AI techniques are generally not now being used in these areas. Our findings then are:

Sales for AI products will grow from about $ 180 million in 1984 to about $ 1.5 billion by 1989.

The mechanism for growth will be enhanced performance/price ratio products, and the introduction of less expensive, stripped down version of AI products. Unit sales will increase dramatically, about a factor of 40 in the same time frame, but most of the growth in unit sales will be unbundled software for personal computers.

The AI market projections of over $ 10 billion by 1990 may not be met, despite greatly increased unit sales.

Saturation of user interests, slow infusion of experience in use, and drastic price cutting will all contribute to a $ 1 to $ 2 billion realistic estimate of the five-year potential for AI products.

In terms of sectors of the market, software sales will grow at the expense of hardware and services.

Both hardware and service sectors of the AI market will grow less rapidly than software, for different reasons. Hardware costs will decline due to improved manufacturing and better design. This will help stimulate the market, but sales will be tied to the capability of the system. Hence software itself is more key to increased sales. Services will tend to saturate due to the lack of qualified practitioners, and the elimination, due to in-house or university training, of some of the need for third party assistance.

The market is shifting leadership from hardware and services to software.

As engineering workstations, mainframe-computers on-the-desk and powerful personal computers emerge, the sales of unbundled software will become the largest segment of the market. Due to sheer numbers of host machines, the personal computer based software segment will dominate in unit sales.

Engineering workstations without Symbolic Processors will become increasingly popular hosts for AI software in this decade, but decline later as symbolic processors become less expensive.

General purpose engineering workstations will emerge as a median system of choice when contrasted between expensive symbolic processors and personal computers (Exhibit 5). The general usefulness of these devices in scientific and engineering work will help promote this choice. Add-on unbundled AI software for this market will be particularly popular. As symbolic processors become less expensive and as applications programs grow to a size that the engineering workstation is bogged down, users will shift to symbolic computers. This later trend should begin by 1990. At the low end of the market, 32-bit processor personal computers (such as the Apple Macintosh, and probable IBM entry) will squeeze engineering workstations as a preferred host system.

AI software for personal computers will enjoy popularity, but applications will be initially limited to education and training.

Publishers of AI languages and AI software for personal computer will focus on education and training markets because of the limited functionality and speed of AI software for personal computers. This market will be a lucrative one in this decade.

By 1990 true applications (such as compiled, or "run time" expert system) of AI on the personal computer level should emerge especially in financial, accounting, and retail sales applications. These will quickly migrate to manufacturing. AI will be built into programs such as word processing and spread sheets by that time frame also. These will generally run primarily on personal computers.

EXHIBIT 5

Overview of Product Categories

PRODUCT	AREA DESCRIPTION
AI COMPUTER HARDWARE (includes bundled AI programming language software provided exclusively for each of these computers)	• AI workstation computers with proprietary symbolic processors (LISP machines) • Engineering workstations with 32-bit processors
AI PROGRAMMING SOFTWARE (unbundled programming environments in LISP, Prolog or other AI languages)	• For AI symbolic processor computers • For large mainframes and superminis • For engineering workstation (non-symbolic) • For personal computers
EXPERT SYSTEM DEVELOPMENT TOOLS (includes design shell and tutorials)	• For AI symbolic processor computers • For superminis and mainframe • For engineering workstation (non-symbolic) • For personal computers
NATURAL LANGUAGE INTERFACE (includes design shell and tutorials)	• For AI symbolic processor computers • For superminis and mainframe For engineering workstation (non-symbolic) • For personal computers
EXPERT SYSTEM APPLICATION PACKAGES	• For superminis and mainframe • For engineering workstation (non-symbolic) • For personal computers
SERVICES	• Commercial research • Government • Maintenance

<u>Expert systems and AI applications have the best potential of the major types of unbundled AI software.</u>

Because of the desire of many first time buyers to be up and running quickly, those systems that come closest to being fully capable will sell the most readily. Expert system shells, that allow users to input special knowledge efficiently, and applications that include or access already existing data, fit in this category the best.

Natural language software will be popular for mainframe, and personal computers due to the higher fraction of "novice users".

Time shared mainframe program and personal computer programs interfaced by a natural language shell that allows the novice to ask inaccurate questions to access data bases represent a good market because of the value and efficiency of such services.

Engineering workstation users, who will be better trained in the details of the software will not find such assets useful.

The controversial "battle" between PROLOG and LISP may be decided by an outsider.

Academic researchers continue "holy wars" over allegiance of AI "mother tongue" languages. Entrepreneurial firms are ignoring this and writing AI software in unique new ways that allow both LISP and PROLOG type reasoning to take place. When these are coded in assemly language or C language, the speed of execution can be excellent.

CONCLUSIONS:

Manufacturing use of AI will grow from 10 % to 1/3 - 1/2 by 1990 if:

- Real time control applications emerge
- No AI skill needed to use, install
- It runs fast enough on non-symbolic machine
- Software license costs no more than $ 100.00 per installation

While vendor products exceed manufacturing needs, in terms features, they fall far short in terms of use in plant because:

- High entry cost
- Need for skilled guides
- Training needed
- Experience base, to install lacking

Thus, repackaging/rewriting custom product for mass market is required.

In conclusion, I would like to thank the conference organizers for the opportunity to attend and address this useful and interesting conference.

ARTIFICIAL INTELLIGENCE IN MANUFACTURING
T. Bernold (editor)
Elsevier Science Publishers B.V. (North-Holland)

DESIGN AS A TOOL FOR CORPORATE IMAGE AND COMMUNICATION

Paolo VITI
Ing. C. Olivetti & C. spa
Direzione Corporate Image
12, Via Meravigli
20123 Milano

The subject, product design as communication, may require a brief premise. We do not need to recur to recent developments in semiology and its perhaps too facile, even occasionally frivolous, influence on visual, figurative or plastic art theories, to declare that every human object inevitably bears communication value as sign, as conventional message, as language. Everything that is made, like every attitude, every convention whose repetition becomes form and every departure from such convention, are meaningful signs which may be expressed in the international medium of a language cipher whose key is as universal as the directional arrow along a highway or at the almost unconscious level of the cryptograph sealed into a bottle and cast upon the waves without establishing who the receiver is to be and almost without knowing the sense of the message itself.
But it is not in this too immediate and self-evident sense that we assume product design to have communication value: any design at all must communicate in a general way the presence of a technology, a cultural condition, a social situation, a geographic milieu, a moment in history. But in this present context we mean to use the word communication in a more limited and specific manner, though always in relationship with these wider determinants.

We can say that the communication values we are interested in as far as the industrial product is concerned are those of its use and function then of its relation to other products of the same line or industry and, in a more general way, of its relations to the technological and formal values of the production of a period, of its image content in terms of recognition and characterization not only of the object itself but of the firm which produces it and puts it on the market; of its relationship, too, to other more specific and direct forms of communication employed by the industry to express its nature and personality; finally, of the broad cultural and social values it implies or declares in the context of the technological and formal culture of its time, its social conditions and the specific situation of the industry as regards the other institutions in which the personality and the traits of the historic moment unfold and take form.

In looking now briefly at these points I shall quite often proceed by example, referring particularly to Olivetti products because they are the ones I am most familiar with and because their specific nature and their technology are such as to make meaning clear: but the general line of the argument would be valid for any object made by human beings.

I want to begin a trite observation. If it is true that everything made communicates something even if the maker is not aware of it, as Molière's bourgeois gentleman spoke prose without ever suspecting it, still what we mean to talk about here is conscious communication: we are interested in those values of international communication that the object presents and by which it is distinguished in a design concept neither wilfully misleading nor approximative. In this sense any piece of equipment bearing signs - a handle, a chair, a pre-formed component - must in the first instance communicate its function and use with maximum clarity and evidence. Let us consider a typewriter or a machine for data processing or calculation. The focal point of such a machine is the keyboard: it is here that the operator's maximum concentration and tension is centered. It is the keyboard, the shape of the keys, the graphic clarity of the signs (whether letters or numbers), the colours that distinguish operational signals, the impression of ease, simplicity, workability that suggest, that offer the fundamental communication to the operator.
True design is, from this point of view, design which tranquillizes, normalizes, put the operator at ease, convinced of his or her command of the instrument. Other levels of communication stem from the form of the object, its dimensions, the way it fits into space, the indications of a man to machine relationship that clearly responds to real experimentation. In concluding, we might say that the rightness of design can be measured by the degree to which a product communicates its nature and its possibilities, reducing ambiguity and uncertainty to a minimum or eliminating them altogether, so that it relates to the user with maximum operative simplicity.

But no object exists alone; it is but one of the many elements in a landscape of objects among which its presence is not a privileged or isolated moment, but one of the articulations of the overall space with its tension and its rythm. To return to our example, the typewriter, the calculator, the data processing unit, are all part of a story that has unfolded over a long period; the industry producing them has projected other generation of simpler - or more complex - machines, with diverse technologies, perfecting and innovating. More than a century has gone by since the first typewriter. At least two technological leaps have intervened: electricity and electronics. Product design must communicate together continuity and innovation. No product is born without parents. It is part of an industry producing dozens or hundreds of other of other products, more traditional or more revolutionary, but linked by the same thematic logic and the same end: to create, elaborate, transmit, broadcast information in writing and in figures; to work with alphabets, numbers, function, reducing the multiplicity of language, the sense of work, the results of management - in sum the products of human creativity - to information visible and documented.
So, the further level of communication the right design must furnish is the unity - may I say the style - of an industry in all its productive complexity.

Not in the rather poor exterior sense of a grammatical, formalistic denominator, but in the truer, "aristocratic", sense of an expressive criteria, a cultural point of view valid for every object made, through which each of these objects reveals its common spirit and a single quality of analysis, linking to those which have gone before it and those which are its contemporaries, so that each stands for an identity, a choice of direction, an expressive morality with which faith is kept. If in the past it was possible to say the style is the man, from this point of view the style is the industry in the measure in which the family or families of objects it produces belong to a common civilization of form by which it is recognized and for which it is perceived, identified and chosen.

But an industry does something more than make the objects that form its catalogue and identify it. Around these objects, supporting and illustrating them, indeed as their condition and premise, there are other elements which propose the nature, the quality and the style of an industry and their communication value, even as they constitute and reveal the "culture" of the industry itself, is as powerfully characterizing as the product itself. Foremost among these physical environment within which the industry lives and operates: its factories, its offices, its sales outlets, its social, cultural and recreational plants, the way in which it fits into the community of which it is a part; but also the sense of its trademark, wherever it appears, the distinctive cut of its publicity, its technical and promotional literature, the graphics of its printed material, the ways in which its production is presented from fair to showroom. And, last but not least, the general level of its behaviour inside and outside its own walls, the way it approaches the universe of possible clients and the vaster field of action to which its economic power gives rise.

Here we are to close to the central point. An industry is not a closed world, feeding upon itself, creating its own separate and different culture, its own solitary language. When it occurs impoverishment is automatic; solitude leads to silence, or at best, to artificiality. In the measure to which an industry is a living organism which can chose its directions and is aware of its ends - that is value-bearing - its participation, osmosis, overlapping with other institutions in the society, is a basic proposition, a condition and a necessity.

An industry that feeds parassitically upon the intellectual, creative and scientific forces of the society in which it operates is a dying industry. And it is already dead when it calls for - or takes up - a relationship of dominance, monopoly, exclusivity: when it acts as mentor to these forces. But if we take vitality as our hypothesis - and it is the only hypothesis useful to our argument - and we see the industry as an organism producing useful goods and therefore continually engaged in research, technology and rationalization - in the non-destructive exploitation of resources - then its position as regards other social institution from the school system to the scientific community and political organs, from the artistic world to the mass media, is at once parallel, competitive and of mutual exchange. Industry's <u>culture</u> can exist only if it is not kept under glass, but lives, stimulated and stimulating, in a live context made up of other industries and of the social institutions in general.

Then product design will not represent simply the level of formal and cultural development of a specific industry but will be a vital part of a whole in which other products, graphic signs, general and specific languages participate; and its communication level will be directly linked to and interdependent with the endless, perennially interchangeable mobile, panorama of objects and products a society expresses and multiplies and devours without pause: effective and individualized as its innovational values, its continually evolving rapport between form and function, its technological accretion - the future charge its form holds and suggests - are in unique, highly defined relationship to the universe of other objects, establishing a moment of tension, proof of evolution or continuity, repetition or revival, the rythm of a creativity constantly at work, or the persistent, known, recall of a tradition. Or all of these together, and with them the infinite shades of possibility between old and new, obsolete and unexpected, succes and error, safety and audacity, in which the human vocation to form expresses and crystallizes itself daily, imposing its tendency to prolong and enlarge through instruments that multiply its capacity, its contact, to act upon the world and transform it. To make the ancient and persistent myth of life as creative force, copy or challenge of divine creation, come true in daily labor.

Though I do not subscribe wholly to the concept that interpretes the universe of what has been made by men coincident with the history of art (George Kubler: "The Shape of Time", 1972), still it is true that what we are used to call art and the object come into being side by side and belong to civilization from which both are inseparable. Though aesthetic thought and critical analisys have assumed, in the history of Western civilization at any rate, a wholly different stance when considering art (in which men express above all their own personality, their inner being and their relationship with existence) and practical, technological production, meant to serve an end (area to which industrial design undoubtedly belongs), history itself seems to show inequivocably that categories like "beauty" and "utility", however distinct in intent, meaning and values are nevertheless born in a common social environment which includes the cultural and moral climate and does not exclude the technological level. Completing one another, both areas mirror and together define the character of a period. If this is so, we can say that there is an ultimate degree of communication in product design, and it is that of witness and partner in the building of its period not only in so far as regards other artifacts and objects created to serve human needs, but within a general formal, figurative - we may even say artistic and intellectual - culture of its times. It is not difficult to see the common citizenship of the industrial products of Le Corbusier or Charles Eames, Gropius, Mies or Nizzoli, for example, and the work of Kandinskij, Klee, Picasso or Mondrian, and we can do so without even adopting the ideological rigor of the Bauhaus group which proposed a sole methodological approach to invention, analysis, formation, teaching and circulation for pure and applied art, architecture, sculpture, painting, graphics - design in all its manifestation. Everything is inter-related, and in a world increasingly unified and informed, characterized by a vital need for rationalization, controlled consumption, ordered space and standards, industry cannot but desire to contribute to the image of its time with products which synthetize the varied meanings we have been talking about and communicate emblematically an idea of the age with its multiplicity of values and the sense of its achievements.

That is why design research must be encouraged, broadened, subjected to constant testing with the same severe perseverance - that may even become heroic single-mindedness - with which an artist pursues his formal research and the same discipline, severity and moral engagement with which the scientist proposes new hypotheses and tries out a more advanced technology. The object is, of course, ephimerous, destined to extinguish itself as progress passes on, while the work of art is made for time, if not eternity. Yet though limited by its briefer life-span it remains, beyond its function and the practical use for which it was born, always and necessarily form; it represents a spiritual value, an organizing will; the realization of an idea of spatial order: on its surface an infinitesimal fragment of civilization sparkles and that civilization is also determined by it, assumes features, may at time evolve and be realized. In a word, like any human act which participates of consciousness, it may, and so it must, win the inertia of habit and banality, and take its place among the things that are made because they are worthy to be imagined and brought into being. An object too has a destiny and a life and of it, as of a human being, the worst one can say is that its life is useless, without meaning and without character.

ARTIFICIAL INTELLIGENCE IN MANUFACTURING
T. Bernold (editor)
Elsevier Science Publishers B.V. (North-Holland)

CHAIRMAN'S SUMMARY

Patricia A MacConaill

Commission of the European Communities
Information Technologies and Telecommunications Task Force

AI in CIM : the Motivation

Why is manufacturing a target area for Artificial Intelligence? Is it simply that AI people regard robots as sexy, or is it something more pragmatic? The latter seems to be the case, with speakers describing several aspects of manufacturing where further progress depended on the availability of new techniques of information processing. These included:

- fast processing of data from multiple sensors;
- smart scheduling to cope with increased complexity;
- on-line diagnostic techniques, to prevent costly errors in production.

Rate of Application

Several speakers examined the state of the art of Artificial Intelligence in manufacturing, concluding that, despite vigorous activity in the universities, there was relatively little penetration of AI into the real production environment. Some activity was reported in scheduling, ie, decision support for scheduling activities. However, such systems as exist are not yet closely integrated within the manufacturing environment. Knowledge-based systems were reported addressing areas close to manufacturing, e.g. configuration design.

A point made very strongly by most of the speakers is that the acquisition of knowledge by the knowledge-based system is a non-trivial problem, and that highly trained specialists would be needed to organize knowledge acquisition on any practical scale. These knowledge engineers would need to work with the person in whose head the knowledge resides, in order to reach a credible knowledge-based system. On this theme, Professor Yoshikawa gave us a very interesting example (on the theory of design), which illustrated that even the practitioners in a specialised field are not necessarily aware of the methodology and thought processes that they use in carrying out their everyday work.

Limitations on progress

What factors limit the exploitation of AI in manufacturing? One speaker commented on conservatism in manufacturing; it is a "show me how" environment, where there is a strong reluctance to trying new techniques before their success is proven. At the moment there is a lack of an installed base of KBS. Another said there was a critical mass limitation, and a shortage of what Mark Fox calls 'masters'.

Cost is also a limiting factor: Dr Knasel said that in order to penetrate the market to any degree, knowledge-based systems would have to be repackaged for the mass market, because otherwise they will not be cheap enough.

Corporate attitude and management considerations

We had many interesting comments, and quite a few clashes (that is what keeps people awake) about the views of the corporation, and the views of management. There seemed to be agreement, however, that we need to master advanced information techniques, and that a key position in the organisation of the future will be that of the information manager. This trend may already be observed in certain companies.

It was pointed out by Dr Stotko that managers are subject to a lot of stress, they are overloaded, their diaries are chaotic, their secretaries won't speak to them. I can say this from the heart, because I have reached the stage where even my children won't speak to me! Another speaker pointed out that knowledge-based systems could eliminate several levels of middle management, which are only there because of the general chaos in organizations.

Societal Factors

One speaker mentioned that in Europe, concern for the individual and a higly unionised workforce can impede the rate of progress in automation in comparison with some other parts of the world.

Education and training

Everyone agreed that the modern worker needs vocational training, not just when he or she emerges from the educational system at age 18 or 21 but throughout life, so as or she is adaptable to new technologies. At present this is far from the case. Professor Hirsch gave as an example the Tornado, a typical aerospace product with a life cycle many times shorter than that of a worker. At the end of the Tornado life cycle, the highly skilled workforce was discarded. This is very unsatisfactory - a waste of people, and a waste of skills. I think we would all agree that the educational system should be adjusted to the worker's life cycle, so as to enable him or her to adapt his or her skills to new technologies as and when necessary.

Conclusions

Those whom I met at coffee time know that very few people could be more prejudiced than I on the subject of AI: I was quite honest about this to Dr Bernold, when he invited me to chair this conference!

However, this symposium has been most instructive for me, it has tended to reduce my prejudice. One of the most reassuring things that Professor Fox said was "The AI fraternity recognizes Murphy's Law". Certainly one had the impression that his team had begun to comprehend the complexity of manufacturing, and the scope for disasters! They were exploiting the techniques of operational research, and looking at the early history of computing, where we discovered that there were some very good ways and some very bad ways to solve problems.

On the side of manufacturers, there will remain a healthy scepticism - a demand to be convinced. Bernd Hirsch, when I referred to him as a 'master'(of manufacturing), said "No, I am only an apprentice". He said that whatever his achievments, he always felt he could have done better if he were given another chance. I am sure many of us feel this frustration in our working lifetimes. There is never time or energy enough to refine and improve our achievements. The fear has been expressed by journalists and others from the floor that AI is going to make us all redundant. I don't agree; I don't think that AI is going to displace human effort, I think it is going to help us to do things better.
I have found this conference extremely stimulating. I very much enjoyed listening to all the speakers and I would like to thank the organizers and the support staff, the interpreters and, last but not least, the audience, for a very interesting and exciting event.

APPLICATIONS

ADVANCED ROBOTICS

ARTIFICIAL INTELLIGENCE IN MANUFACTURING
T. Bernold (editor)
Elsevier Science Publishers B.V. (North-Holland)

Machine Vision Applications of Associative Networks

J. Michael Oyster

Walter Broadwell

Fernando Vicuna

Gary Silverman

Robotics Project
Los Angeles Scientific Center
11601 Wilshire Blvd.
Los Angeles, CA 90025
USA

ABSTRACT

The Associative Network, developed by Claude Cruz-Young and his associates at the IBM Palo Alto Scientific Center, see (Cruz 1982), is a computer architecture emphasizing local and asynchronous control, high parallelism and redundancy. The Robotics Project at the IBM Los Angeles Scientific Center has been exploring the use of Associative Networks to solve complex machine vision problems.

Two very important characteristics of an architecture for parallel processing are fault tolerance and the communications protocol between processing elements. Fault tolerance is important because with a very large number of processing elements the probability of an element failing is high. Similarly, deadlock arising when many processors simultaneously require access to a communications link is often a bottleneck in parallel computing architectures. The Associative Network is fault tolerant and avoids the problem of communication deadlock.

One of the advantages of the Associative Network is the flexibility with which connections can be defined. Other parallel processing schemes, such as neighborhood processors, cannot define communication pathways between arbitrary pairs of processors. The flexibility of the Associative Network is essential for extracting global information from an image.

This paper presents research applications of the Associative Network to common problems in machine vision. These research applications include: edge-enhanced images and the Hough transform. A longer version (Oyster, et. al. 1986) has in addition: boolean operations, convolution integrals, Gaussian filtered images, region-filling algorithms and pattern recognition. Each of these computations require a distinct approach to Associative Network design, demonstrating their versatility. Our research has shown how to use Associative Networks to exploit the natural parallelism in a variety of image processing and machine vision problems.

1.0 INFORMAL DESCRIPTION OF ASSOCIATIVE NETWORKS

This paper is much abbreviated. For further details, see (Oyster, et. al. 1986).

In the traditional digital computer, collections of binary numbers represent the program and data. The precision of the binary representation leads to great stability and reproducibility. Largely for this reason, digital computers have nearly displaced analog computers over the last 15 years. This precise binary representation is not very flexible. Small changes in the data or program, caused either by hardware errors or errors in data collection, can cause sudden changes in the behavior of the computing system (e.g. system failure). A more flexible approach to computation can be advantageous.

Associative Networks, originally developed by Claude Cruz-Young (Cruz 1982), can be made relatively insensitive to changes in the program or data. This is because the evolution in time of the Associative Network is a voting or averaging process. Different processing elements can compete and cooperate to produce a global collection of states insensitive to the accuracy of a single processor or a single piece of data. The state of each processing element is a continuous value determined by a weighted average of the states of neighboring processing elements. The Associative Network is in principle a highly fault-tolerant architecture.

The Associative Network offers a simple and elegant solution to the communication deadlock problem in parallel computing architectures. Computing the state of a processing element as a weighted average of its neighbors *defines* the processing of multiple inputs by each node in the network. By definition, there is no deadlock.

As the Associative Network differs greatly from conventional computing architectures, a new approach to programming is required. The execution of the Associative Network depends on the topology of the network and on the weighting factors (or link weights) assigned to each pair of connected nodes. For the Associative Network programmer, the task is to specify the topology of the Associative Network, the connecting weights, and the node parameters. The topological definition of the Associative Network contrasts conventional programs containing sequences of instructions to be executed in series or in parallel.

The Associative Network is a collection of nodes connected by directed edges or links. Associated with each link in the network is a real number (positive or negative) called the link weight. Associated with each node is a real number in the interval [0,1], called the "activity" level of the node. The network is updated in a series of discrete time steps as dictated by the Network Update Algorithm. The "activity" or "state" of each node depends on the activity levels of adjacent nodes in the network in the preceding time step. The state of each node, i, depends on the activity levels of only those nodes, j, in the network such that a link points from j to i, and on the link weight associated with the link from node j to node i.

Figure 1 is a simple example of an Associative Network. The diagram on the left is the initial state of the network and the diagram on the right is the next state of the network. In the initial state, Node A has an activity of 0.1, Node B has an activity of 0.2, and Node C has an activity of 0.3. The respective connections have weights of -0.5, 0.3, and 1. In the following time step, the state (or activity) of Node D is:

$$-0.5*0.1 + 0.2*0.3 + 0.3*1 = 0.31.$$

The Associative Network is not updated with a *true* weighted average. One disadvantage of the weighted average is it can be unbounded. If we allow the node values (or activity levels) to be unbounded, then the network can become unstable, overflow, or fail to converge. Thus, we restrict the node values to fall between 0 and 1. If the computed average is between 0 and 1, then a true weighted average is calculated. If the computed average is greater than 1, then it is truncated to 1. If the computed average is less than 0, then it is forced to 0.

In addition, the Associative Network performs thresholding. One use of the threshold is to eliminate noise due to weak signals. The threshold can also be used to provide various logic and control characteristics to a given node in an Associative Network. Detailed examples making good use of the threshold are given (Oyster, et. al. 1986).

After thresholding the weighted average, a gain factor can be applied. Since the gain factor is applied before the saturation limit is applied (ie. the maximum value of 1), setting a small gain factor can prevent a node from going into saturation, and setting a large gain factor can force a node to saturate in the presence of a weak signal.

Figures 2(a) through 2(c) demonstrate threshold, gain, and saturation. In Figure 2(a), the node values for A and B are averaged (using link weights of 1 for simplicity) to obtain the value 0.6 + 0.25 = 0.85. After applying the threshold of 0.5 the activity for Node C becomes 0.85 - 0.5 = 0.35. In Figure 2(b), the threshold is raised to 0.8 and the gain to 100. The activity for Node C is then min(1, 100*(0.85 - 0.8)) = min(1, 5) = 1. This illustrates the gain and saturation. In Figure 2(c), the threshold is raised to 0.9. The activity of Node C is 100*(max(0, 0.85 - 0.9)) = 100 * max(0, -0.05) = 100 * 0 = 0. This illustrates the requirement that node values must be non-negative. Observe that the negative value is forced to zero *before* the gain is applied.

The properties informally described here determine the Associative Network's processing power for a fixed point in time. In addition, the Associative Network has the power to reconfigure itself dynamically without reloading or recompilation. The topology of the connec-

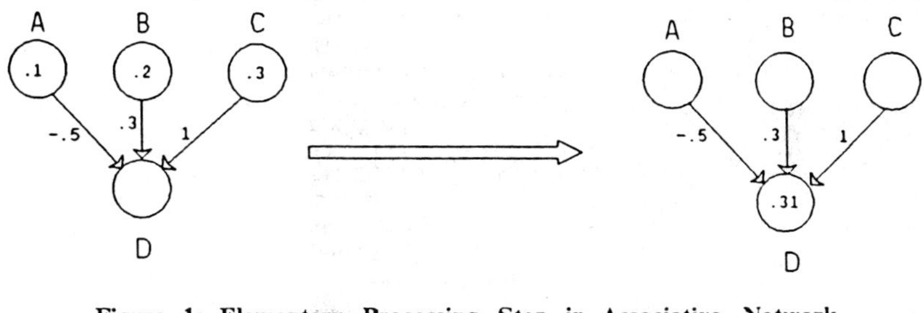

Figure 1: Elementary Processing Step in Associative Network

tions between nodes does not vary with time, but the link weights between nodes (which determine the "program" of the network) are variable [1].

---

1 The *effective* topology of an Associative Network may change over time. Paths that are traversed frequently can be reinforced, while others may effectively disappear or "dissolve" (i.e. stop allowing flow of activation over them).

Link weights are updated by the following heuristic:

- The change in link weight is the product of the activity levels of the two connected nodes times the network plasticity factor.

The network plasticity is a constant determining the rate of network adaptation. In addition, there is a correction term allowing for the decay of inactive parts of the network. For simplicity, these are ignored in the present discussion.

At first glance, the Associative Network may seem very limited. Its design is very simple, and it works with continuous rather than discrete values. How can it possibly make the logical decisions necessary for the control of complex programs? As shown in (Oyster, et. al. 1986), the Associative Network is capable of performing the logical operations: "AND", "OR" and "XOR". Since these three operations form a universal set, the Associative Network can compute any multivariate Boolean function. Utilizing Boolean functions, it is possible to implement the elaborate control schemes necessary for useful computation.

The Associative Network is capable of performing tasks routinely performed by conventional computers. More importantly, it has the added features of massive parallelism, fault-tolerance, and self-adaptation. We demonstrate these capabilities with a few applications to machine vision.

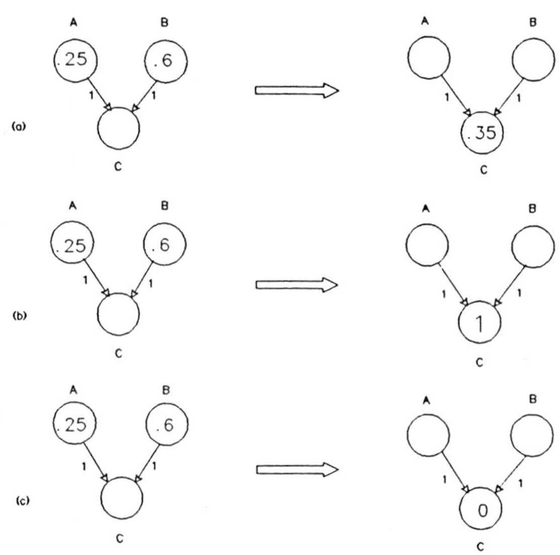

Figure 2: Examples of Associative Networks with Gain and Threshold

(a) Threshold = 0.5, Gain = 1

(b) Threshold = 0.8, Gain = 100

(c) Threshold = 0.9, Gain = 100

2.0 MACHINE VISION APPLICATIONS

2.1 Low-Level Edge Detection in Associative Networks

Low-level edge detection determines the locations of those pixels in the image plane lying on the boundary between two regions. Low-level edge detection does not determine either the geometry of the edge (eg. line, circle, ellipse) or its topology (eg. closed, connected). Higher level processing is required for this purpose.

Edge detection is but one element in a particular vision hierarchy. Not all machine vision schemes require edge detection. It is a frequently used intermediate step to vision. Marr (1982) points to experimental evidence indicating that edge detection plays an important role in animal vision systems. It is a stepping stone to obtaining line segments, vertices, corners and higher level geometric objects.

Operators for low-level edge detection include the Robert's edge operator, the Sobel operator, and the Laplacian of a Gaussian (LOG) operator. Each of these operators is a special case of the convolution. As is shown in Oyster, et. al. 1986, any convolution can be computed by an Associative Network. Thus, the standard edge detection operators can be implemented with an Associative Network. Cruz (1983) demonstrated that the Associative Network can perform edge detection using the LOG operator. We have also chosen the LOG operator because one can adjust the filtering properties of the LOG to accommodate different kinds of boundaries, especially weak boundaries spread out over several pixel diameters. Our work extends the work by Cruz-Young and demonstrates the systematic application of the primitive operations defined in Oyster, et. al. 1986.

The edge detection is performed with six node plane transformations. The transformations are illustrated in Figure 3. GF is the Laplacian of a Gaussian transformation. LE, RE, TE, and BE detect left, right, top, and bottom edges, respectively. OR combines the results of each of the edge detectors into a single composite outline.

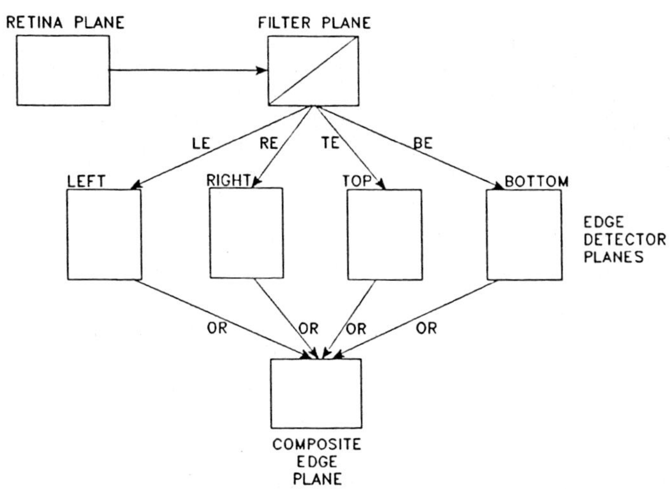

Figure 3: Node Plane Transformations for Edge Detection

GF is defined by convoluting the LOG with the retina plane. The slash in the symbol for the plane is an abbreviation for two planes of nodes acting together as a single plane with values defined in [-1,1]. Oyster, et. al 1986 shows how the connections are defined for an arbitrary convolution.

Figure 4 provides a one-dimensional slice of the left-edge detector. For each pair of split nodes adjacent in the x direction, the "AND" operation is performed to detect a positively saturated split node located one node to the left of a negatively saturated split node. The plane operators RE, TE, and BE are defined similarly.

The edge detection algorithm was tested using a variety of Gaussian widths and test images. If the width of the Gaussian is greater than one, then the edge detector must be slightly modified.

2.2 Associative Network Hough Transform for Detecting Lines

In this section, we define a discrete version of the Hough transform and show how to implement it as an Associative Network. Consider a normalized coordinate space S defined over the variables i and j. Let i and j range from -1 to 1 and take on the values -1= -N/N, -(N-1)/N, -(N-2)/N, ... , (N-1)/N, N/N=1, where N is an arbitrary positive integer. Let i and j be indices for the discrete pixel plane of dimensions (2N+1) x (2N+1). Let the value of each pixel, P(i,j), be the pixel intensity of the image to be transformed.

Now consider two additional planes of pixels S1 and S2, each isomorphic to S. Call the index variables i1,j1 and i2,j2 for S1 and S2, respectively. Map P(i,j) to each pixel closest to the lines with equations, j1 = -i*i1 + j and j2 = -j*i2 + i, in the respective planes S1 and S2. More precisely, P(i,j) is mapped to the pixels on the boundaries of pixel planes S1 and S2 closest to the end points of the two lines. The endpoints are then connected using the Bresenham line drawing algorithm (Foley and Van Dam 1984). Plane S1 is for detecting lines of slope between -1 and 1 and Plane S2 is for detecting lines of slope less than -1 or greater than 1. Vertical lines can be detected by S2 since the variables i and j have been permuted.

We now implement the transform in an Associative Network. For each of the three planes assign a node to each pixel. For every node n(i,j) in the image plane define connections of weight α to each node in the transform plane in the image of the discrete Hough transform. For example, consider the 5x5 array of nodes in Figure 5. We show here the connections from the node (i,j) = (1/2,1). For greater readability the connected nodes are marked with an "X". The diagram illustrates the only the connections. Stimulus would first have to be applied for the nodes marked with an "X" to become active.

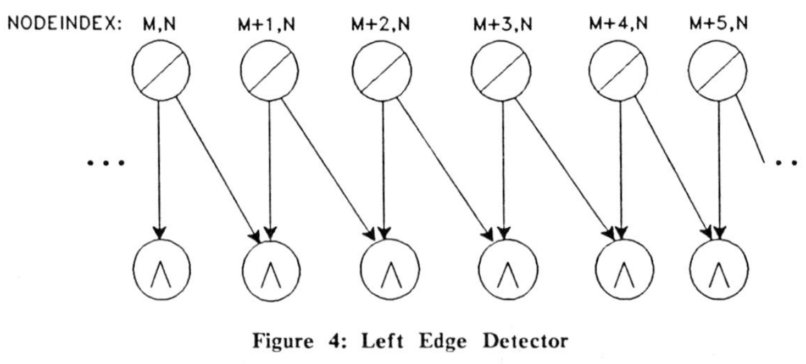

Figure 4: Left Edge Detector

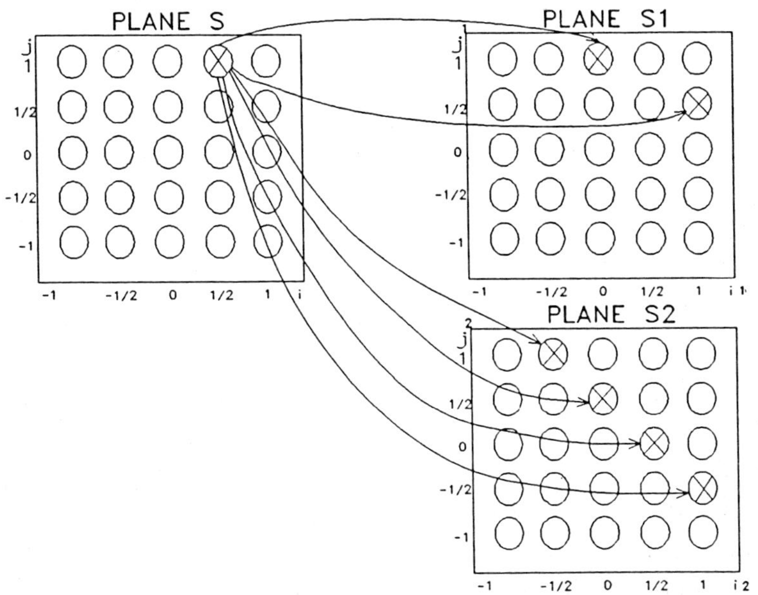

Figure 5: Representative Connections for the Hough Transform.

The reader will note that the number of connections increases rapidly as the resolution of the image is increased. For a 64x64 pixel image, approximately 64 connections will be needed for each node. Human brain cells can have up to 10000 connections, so in biological systems this does not appear to be a serious limitation.

Each node in the transform space represents a line in the original image. Thus the intersection of two lines in the transform space also represents a line (defined by at least two points) in the original image. Lines in the original image can be detected by finding nodes in the transform space belonging to at least two lines in the transform space (or equivalently in the image under the Hough transform of a least two active points in the image node plane). Intersections of lines in transform space can be detected by setting a threshold in the transform space such that a node is not active unless it is receiving activation from at least two nodes in the original image. Given a threshold of 1/2, a gain of 2, and connections of weight $\alpha = 1/2$ in the transform plane; lines in the image can be detected. Note that the link weight, threshold, and gain are taken directly from the "AND" macro defined in Oyster, et. al. 1986.

BIBLIOGRAPHY

Cruz, C.A.(1982). *Associative Networks.* IBM Palo Alto Scientific Center Report No. G320-3474.

Cruz, C.A.,& Meyers, H.J.(1983). *Associative Networks II.* IBM Palo Alto Scientific Center Report No. G320-3446.

Foley, J.D.,& Van Dam, A.(1984). *Fundamentals of Interactive Computer Graphics.* Addison-Wesley

Hough, P.V.C. (1962). *Method and Means for Recognizing Complex Patterns.* U.S. Patent 3069654, December 18, 1962.

Marr, D.(1982). *Vision.* W.H. Freeman and Co.

Minsky, M.,& Papert S. (1972). *Perceptrons.* MIT Press.

Oyster, J.M., W.B. Broadwell, F. Vicuna, "Associative Network Applications to Robot Vision", IBM Corp., Los Angeles Scientific Center Report G320-2777, February, 1986.

Rosenblatt, F.(1961). *Principles of Neurodynamics: Perceptrons and the Theory of Brain Mechanism.* Washington, D.C. Spartan.

Sabbah, D.(1985). Computing with Connections in Visual Recognition of Origami Objects. *Cognitive Science*,9, p25.

Skrzypek, J.(1985). Computing Architecture for Spatial Preprocessing of Visual Information. *SPIE Proceedings on Robots and Vision* 1985, Cambridge MA.

Skrzypek, J.(1985). Computing Architecture for Temporal Preprocessing of Visual Information. *SPIE Proceedings on Robots and Vision* 1985, Cambridge MA.

ARTIFICIAL INTELLIGENCE IN MANUFACTURING
T. Bernold (editor)
Elsevier Science Publishers B.V. (North-Holland)

THE INDUSTRIAL ROBOTS IN A FLEXIBLE MANUFACTURING SYSTEM: STATE OF THE ART AND PROSPECTS

Dr.-Ing. R.D. Schraft

Fraunhofer-Institut für Produktionstechnik und Automatisierung (IPA)
Stuttgart, Germany

INDUSTRIAL ROBOTS-STATE OF THE ART AND FORESEEABLE DEVELOPMENT

Industrial robots have been around since 1965. They first came into use on an appreciable scale in Europe in the early '70s and by the end of 1983 there were about 13,700 robots on the job. Their major duties may be divided into the handling of tools and handling of work. Surveying current robotics applications in West Germany, it is found that roughly every second unit is installed in the automobile industry. Other than their now classic employment in spot welding, machine loading and arc welding, they are expected to enter a wide range of new uses in future, including machining of parts, loading of machines, and assembly of parts.

WEST GERMAN FIGURES

A review of the industrial robot market in West Germany shows some 200 different types being offered by 80 companies as of the end of 1984. This is a significant figure considering that the three biggest manufacturers and suppliers in the country already command 50 percent of the national market. The first dozen companies to enter the field are already meeting over 80 percent of the demand. Robotics is certainly a growth market and holds apportunities for small and medium firms too. In some fields of application definite market leaders have meanwhile come to the fore whereas others, for instance the high-growth area of assembly operations, are still wide open. Robotics cannot and must not be the cure-all for automation problems of the future, but it is steadily gaining ground in industrial production.

INTERNATIONAL STATUS

Research and development in robotics has had a strong influence on the utilization of industrial robots. The U.S.A. and Japan started into the fundamentals much early than did Germany and the rest of Europe. Differing definitions apart, Japan uncontestedly heads the world in numbers of industrial robots at work (Fig. 1).

At the end of 1983, there were 16,500 robots in service in Japan compared to 4,800 in West Germany, giving the Japanese a clear lead over both America with its maximum number of 8,000 and Europe with 9,000. Latest trends reported from Japan point to substantial increases in the field of assembly.

Absolute numbers of industrial robots in use do not say much about how active an industrial nation is in innovation. A better indicator is the density of the units (number installed per 10,000 manufacturing employees; Fig. 2). On this scale, Sweden lies far ahead of Japan and West Germany. The reasons for this are the extensive social services network in force in Sweden, high labor costs there and the existence of prospective users (a substantial automotive industry). Moreover, efficient robots came onto the Scandinavian market at a very early date.

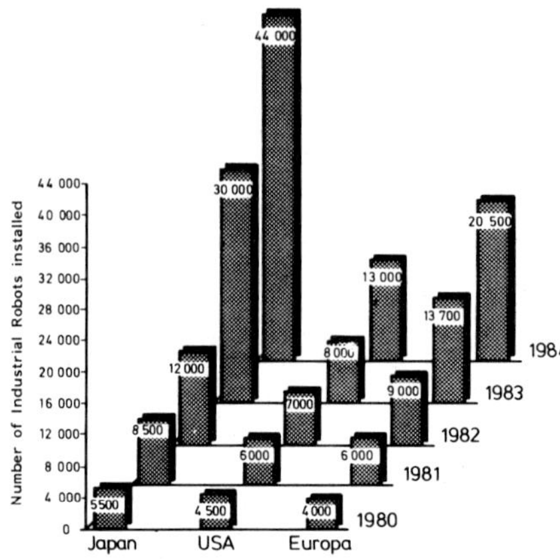

Figure 1
International Distribution of industrial robots in use

	Japan	Sweden	Federal Republic of Germany	France	USA	Italy	Great Britain
No. of employees (million) 1) State: 1983	11,52	0,984	8,601	5,269	21,817	5,388	6,4
Industrial robots installed State: end 1984	44 000 3)	2 400 2)	6 600 4)	3 380 5)	13 000 6)	2 700 7)	2623 6)

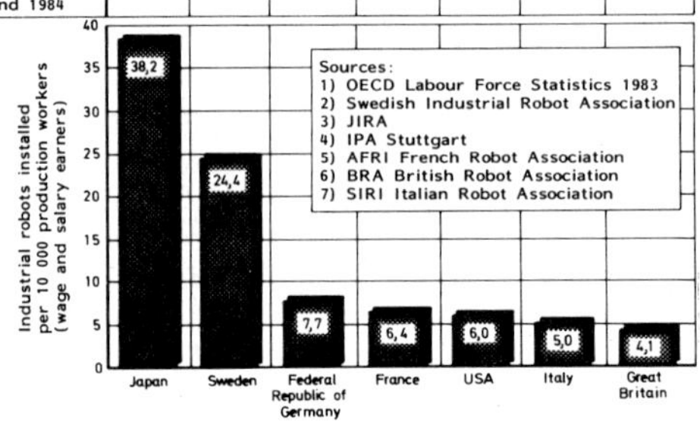

Figure 2
"Robot-Density" in some industrial countries

APPLICATIONS

Industrial robots can be differentiated by those to handle tools and those to handle work. When equipped with gripper arms or tool changers, they can serve both functions.

Developments in the separate applications have differed considerably. The present status and movements over the past four years are illustrated in Fig. 3. They may be commented as follows.

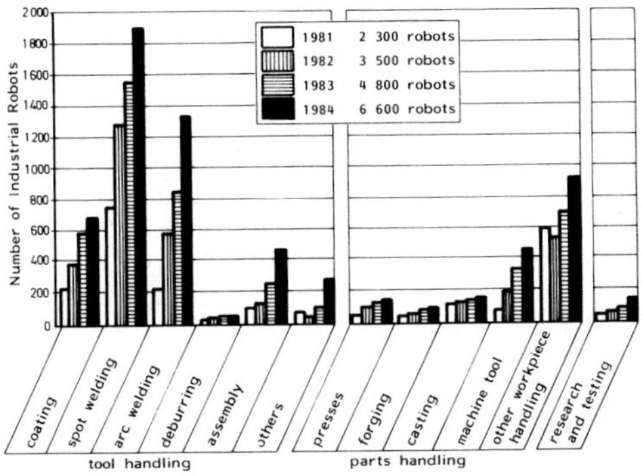

Figure 3
Industrial robots in use in West Germany

Coating
No major technical problems remain to be solved. Economy is the sole yardstick for adoption.

Spot welding
Very strong expansion in use of industrial robots for spot welding is attributed to the accumulated demand in the automobile industry. Growth cannot be expected to continue at the same rate, as the industry is now nearing saturation.

Continuous welding
By and large a great future is predicted for continuous welding as soon as answers are found to as-yet unsolved problems such as weld line tracking, compensation of jigging tolerances and the like.

Deburring
Deburring and machining by industrial robots is looking far beyond the current situation, where huge technical difficulties must first be overcome.

Assembly
In a Delphi poll taken among manufacturing companies, the highest growth rates were generally predicted for industrial robots in assembly of parts (Fig. 4). Conditions are promising for this to be a successful application both as regards product design and the equipment technology.

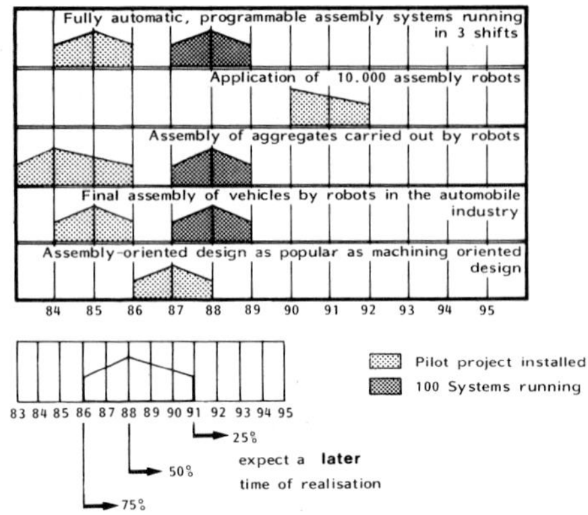

Figure 4
Delphi poll on flexible assembly automation part

Tool handling
On principle, this problem has been solved. Above all, loading and unloading of machines will attain economic significance in the years ahead in order to enable capital-intensive production facilities to operate multiple shifts. Fig. 5 presents an updated forecast of changes anticipated in the U.S.A. for the application of industrial robots over the next ten years. It is not to be expected that like relationships can be assumed for Europe, as a comparison of figures up to 1981 reveals substantial similarity.

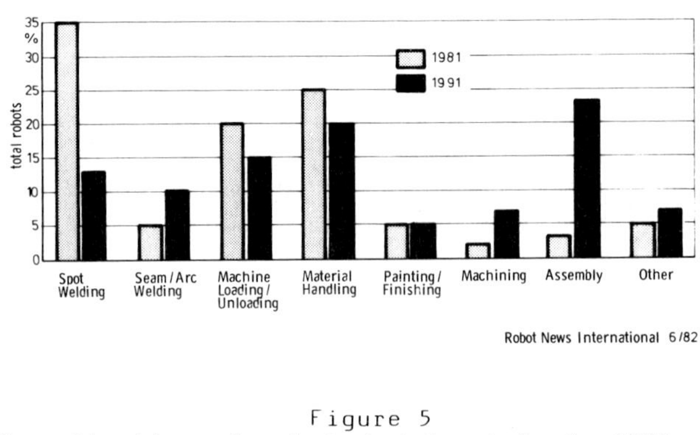

Figure 5
Comparison of applications for industrial robots in 1981 with those anticipated for 1991 in the U.S.A.

Past experience shows the costs of engineering and peripherals and the costs of the robot to be approximately equal.

Today's users are usually not interested in buying industrial robots as automation elements. What is rather wanted is overall system solutions.

With exchangeable hands, industrial robots are able to tackle complex jobs. Different types of work as well as tools can be handled by a robot worker.

TRENDS IN DEVELOPMENT

Developments to be anticipated in robotics may be summarized thus:

- Sensor technology will advance at a fast pace enabling the robots to approach their given task with greater sensitivity and intelligence.
- Robot hands will attain more infinite capabilities. Coordination between hand and eye will become possible in conjunction with sensors.
- Programming of industrial robots will become simpler by other programming methods (CAD/CAM, voice input, higher programming language, semiautomatic programmers and more of the same kind). This will cut the time and expense of developing programs appreciably.
- Robots will be able to work over a larger range of space; in other words they are becoming mobile.
- Endowed with a higher level of intelligence, robots will be able to work in environments that do not require optimum preparation. That, in turn, helps to decrease the expense and effort for developing problem-specific peripherals.
- In the medium term, prices of industrial robots are not due to undergo much change. Despite the availability of controls that perform better and cost less, the price of the equipment as a whole will remain relatively stable for the time being, as the mechanical construction is not getting appreciably cheaper.

ASSEMBLY OPERATIONS

Compared with operations in industrial manufacture, the application of robotics to assembling parts has sooner been laging, although this is the area where the biggest potential for use is seen to lie [1].

While unit labor cost in the manufacture of parts have been brought down a long way by new materials, simplification of products, numerical control of machines and new production technologies, the reverse has occurred in assembling the same parts do the final product. Among other things, the extent to which assembly of parts can be automated will strongly determine the competitiveness of German industry.

Current market demands characterizing the situation in production assembly are:

- Increasing numbers of different versions and models
- Small batch runs
- shorter times in production

Automation of assembly can only take place through more flexible assembly systems on other words. More flexible assembly systems are moreover needed to preserve the existing high level of automation in high-volume production over the long term. In this connection, high hopes are placed in assembly robots as the principal element in new flexible assembly systems.

ASSEMBLY AUTOMATION SYSTEMS

Automated systems for assembly cover a range involving parts weighing from a few grams to a thousand kilograms. Cycle times extend from less than a second for small parts to several minutes for large ones.

Fig. 6 tabulates the applications for automation aids in parts assembly. The main distinctions are cycle time, number of product versions to be assembled, and volume of production. Cycle times are governed by the annual output, with a certain minimum being set by the weight to be handled and size of the parts.

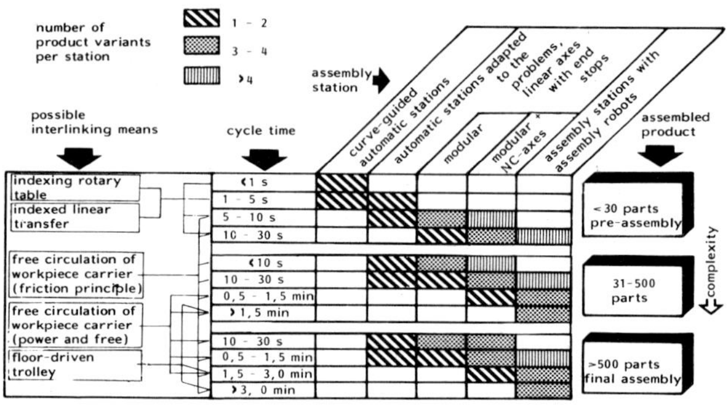

Figure 6
Applications of automation aids in assembly

Large and heavy parts cannot be assembled in randomly short cycle times. Complexity is used as a term to describe the total number of separate parts in the assembled product, but not meaning that all of them are to be assembled by the one system. Products composed of more than 500 parts can be finally assembled from a small number of subassemblies. Many companies distinguish between preassembly and final assembly within the one assembly set-up even. Division of overall assembly is made according to the total number of parts in the final product, i.e., its complexity [2].

The three categories thus obtained are:

- o Products composed of 30 parts or less manufactured in large numbers with short cycle times and in typical "preassembly" operations. They include relays, spark plugs and switches.
- o Products of 30 to 500 parts are assembled in combinations of separate parts and previously made subassemblies. Typical for this category are coffeemakers, pocket calculators, transmissions.
- o Products having more than 500 parts are representative of the "final assembly" category. They differ from the others in being mainly assembled from subassemblies together with the individual parts to fasten them in place. Cycle times are considerably longer than for the typical pre-assembly operations. Joining is a more intricate task. Among these products are television sets and cars.

Design of the assembly stations themselves and the transfer facilities between them is determined by these principal distinctions. For automatic assembly stations only transfer facilities employing indexing of work carriers should be used. Free rotation of the work is preferable to a rigidly timed intermittent feed system in the interests of preventing overall availability from being sharply diminished by minor trouble at any one station.

Pre-assembly (products comprising up to 30 parts), allowing assembly of several versions in cycle times of 10 to 30 seconds, constitutes a highly promising application for assembly robots. Economical feeding and handling of parts is sooner to be achieved in this case than when subassemblies are to be joined, as they usually have to be magazine-fed. Compared to final assembly, the number of possible versions produced is generally smaller in pre-assembly operations.

COMPONENTS OF PROGRAMMABLE ASSEMBLY SYSTEMS

The main components from which programmable assembly stations can be constructed are pictured in Fig. 7. These can be combined with manual workstations and single-purpose automatic stations. Simple products can however be assembled completely by a robot designed for the job.

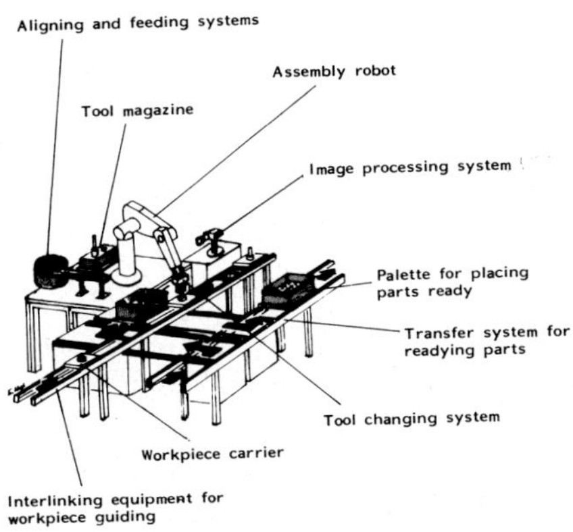

Figure 7
Components for programmable assembly stations in assembly lines

As the main component, the assembly robot has the advantage of

- o free programmability and adaptability to different assembly tasks,
- o small planning and engineering effort,
- o permitting use of sensors.

Manufacturers of assembly robots in West Germany and the models they supply are listed in Fig. 8. The machines are suitable for assembly work by reason of their control and programming language, power transmission and accuracy. There are no hard and fast boundaries; other industrial robots may also be used for assembly, under certain conditions. Serveral German manufacturers are already supplying assembly robots with a horizontal bending arm working to the SCARA principle (selective compliance assembly robot arm), a Japanese development.

manufacturer/ licenser price in DM	model	axes	kinematics main axes	drive	working area	rated load	repetition/ positioning accuracy	machine type	control/programming language
Jungheinrich 124,000	R 55/R 56	5 to 6	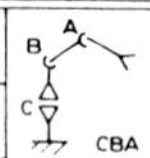	DC servomotors		30 N	± 0,2 mm	vertical bending arm	PTP/CP Rolf
Mantec 99,000	R 3	6		DC servomotors		100 N	± 0,05 mm	vertical bending arm	PTP/CP function keys (Siemens)
FWM/AEG 140,000	Servator 02	3 to 6	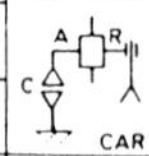	DC servomotors		75 N	± 0,05 mm	horizontal bending arm	PTP/CP function keys (AEG)
Messma Kelch 80,000	Moro 300 + 4	4		DC servomotors		50 N	± 0,02 mm	horizontal bending arm	PTP/CP function keys (IBH)
Bosch	FMS Module	1 to 4	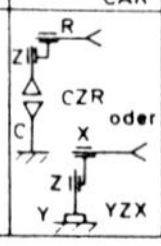	DC servomotors	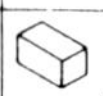 oder 			modular combination of axes modules	

Figure 8
Manufacturers of assembly robots in West Germany and the models they supply
(Status: August 1983)

Two work areas for assembly robots will now be explored in more detail; assembly of automotive and assembly of small parts.

ROBOTS FOR ASSEMBLY OF UNITS

Apart from the flexiblity of assembly robots already mentioned, investigation of the actual assembly operation reveals additional benefits over conventional problem-oriented assembly stations. Faults occurring during the course of assembly can be detected by sensors and corrected by programmed troubleshooting strategies or can signal the need for manual intervention. High flexibility also allows of automating complex joining motions. This will be explained by a programmable assembly unit developed at the Fraunhofer Institute for Production Engineering and Automation (IPA) in Stuttgart, West Germany. Two industrial robots mount a cover, water pump and belt pulley on a car engine, fastening them with a total of ten screws differing in head size.

The cycle for assembly of the water pump is described in Fig. 9. It will be seen how accurately progress of the operation can be monitored by the robot's control system. The water pumps are fed by magazine and must be located on the front end of the engine block without the aid of jigs. This is performed by a gripper-hand robot (Fig. 10). A second robot equipped with a programmable wrench screws the pump in place. The handling robot's motor-driven parallel-finger gripper with programmable opening is visible in Fig. 10. It is resilient in the vertical direction (direction of assembling pump on engine) and has sensors to measure gripping force and correct mounting.

Demonstrated at the 1983 Hanover Fair, this assembly system shows how industrial robots can be employed more extensively in assembly of automotive units. The way existing assembly lines are structured, conditions are favorable for economical use of industrial robots [3].

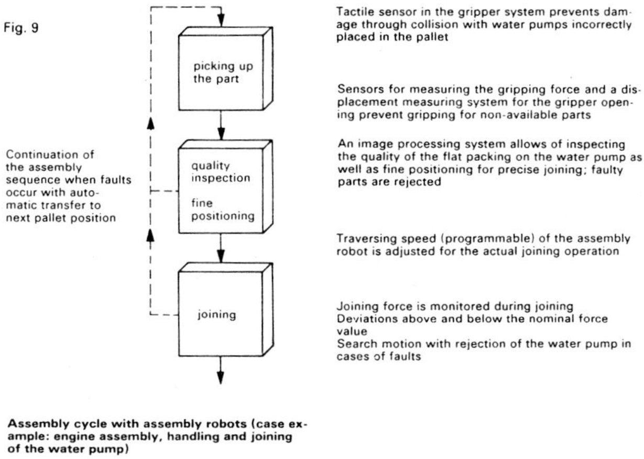

Figure 9
Assembly cycle with assembly robots (case example: engine assembly, handling and joining of the water pump)

Figure 10
Handling robot of the programmable assembly cell with force- and touch-controlled parallel-finger-gripper

HIGH SPEED ROBOTS OF HIGH POSITIONING ACCURACY FOR ASSEMBLY OF SMALL PARTS

These are the demands placed on assembly robotics by the electrical and precision engineering industries. MoRo +4 is a model already offering a tool change wrist and marked by high rate of acceleration and deceleration together with high accuracy of repetition. In the electrical industry, mounting component parts on printed circuit boards is a prominent example of a promising application for small assembly robots.

Design of this robot varies from the SCARA configuration. Instead of a horizontal bending arm, it has a linear-axis arm.

Owing to fabrication of printed circuits already having been extensively automated as far as the standard components are concerned, mounting the specialized components on the board by hand is becoming more and more of a burden. Specialized components in this instance are relays, transformers, chokes, plug strips, trimming potentiometers and the like. Although only some 10 - 20 % of the total, these components cost as much to assemble as the bulk of standard parts. Relatively simple feeding of the parts, including the specialized components, is further reason to view assembly robots as economical for this work. Their use is expected to grow between 30 and 40 percent in the U.S. by the end of the decade.

STATUS OF ASSEMBLY-ORIENTED PRODUCT DESIGN

Rules and guidelines for assembly-oriented product design are formulated with the aim of completing assembly of the product with minimum expense of time, equipment, space and labor [4]. These efforts are supported and actually accelerated by ever newer methods of fastening which enable assembly to be performed more easily, economically and often to be better automated too.

With assembly costs constantly on the rise, designers are increasingly being faced with ever more urgent and concrete demands to design products better oriented to assembly operations. This is in addition to the 15 or more criteria already needing to be considered as production and machining factors (functional fitness, convenience for servicing, suitability for NC machining, etc.). The necessity for assembly-oriented product design derives largely from the fact that rationalization exclusively relating to assembly and not extending to the preceding steps in manufacture is only able to generate a relatively small reduction in overall product costs, no matter how strenuous the efforts are [5, 6, 7, 8, 9].

The problems of assembly-oriented product design were investigated in a study during which over 355 companies were surveyed [10]. Together with numerous other problems, the representative survey inquired into and analyzed the obstacles to automation in product design, possibilities of reducing standard times by assembly-oriented product design, and the instruments available and used to achieve assembly-oriented product design.

Some of the salient results will be reported in the following: At very many companies, insufficient attention is given to the demands for assembly-oriented product design. The most frequently mentioned obstacles to automation in product design are summarized in Fig. 11. According to the respondents, the major obstacles were non-existence of assembly oriented product design in general and poor handling of individual parts.

Specific effects counted on from assembly-oriented product design, beside making automation of assembly possible and easier, include reduction of standard time for assembly. The percentage of companies anticipating just that is given in Fig. 12. Positive effects on these

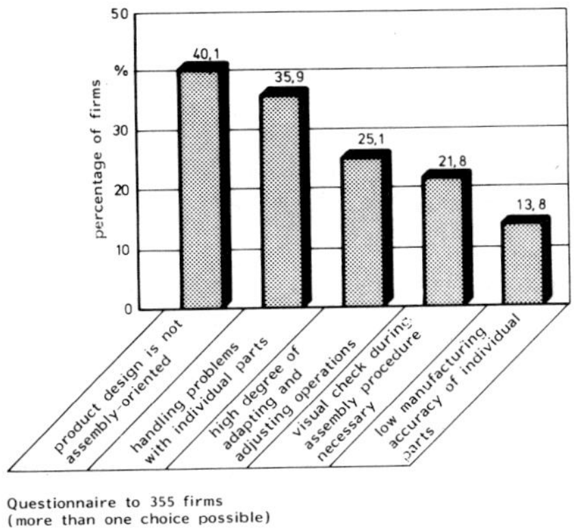

Figure 11
Obstacles to automation in product design

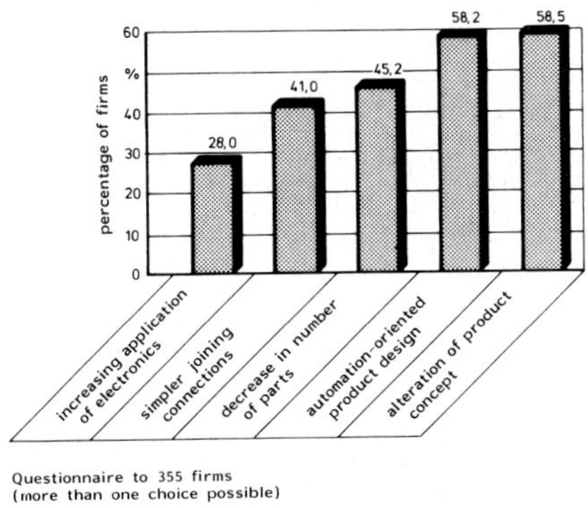

Figure 12
Precentage of firms expecting to reduce the assembly time allowed by means of assembly-oriented product design

times are hoped for from automation-oriented product design and changes in existing product concepts. No small number of manufacturers look to reduction in the number of parts and simplifying the joining of parts in assemblies as a means of cutting down the time that must be allowed for the job.

These high expectations on efforts directed at assembly-oriented product design often fail to materialize, however, because the know-how accumulated on the subject, some documented and some still not, is not always adequately available to the staff concerned or because no systematic approach is pursued to translate the precepts of assembly-oriented product design into actual designs of products [11].

POTENTIAL FOR ASSEMBLY-ORIENTED PRODUCT DESIGN

Design of product

Assembly-oriented product design results in

- o smaller effort for automation of assembly,
- o automation of assembly becoming at all possible, or
- o less manual work as a part of overall assembly operations [12].

According to the level from which planing proceeds, there are various points of entry for assembly-oriented product design, leading in turn to varying levels of rationalization potential [13]. The biggest savings are achieved by adaptation of the overall product make-up, which can only be implemented economically in the long run when designing a new product or model, since the effort to modify an existing design would usually be greater than the economics yielded by the ensuing rationalization. An example of an assembly-oriented overall product design is presented in Fig. 13. It was devised jointly by the product development and assembly engineering departments of the car manufacturer with a view to high-level automation in fabrication by permitting extensively automated pre-assembly of the separate subassemblies which go to make up the complete car. As a procedure for the medium term, design engineering adaptation of subassemblies can be made on existing products with comparatively little expense. When doing so, it is essential to preserve the interfaces to other subassemblies and components as they are. The rationalization potential of this approach is smaller than that to be gained from tackling the product design as a whole. An example is found in Fig. 14. The original 13 parts needing to be assembled in several directions of fitting have been reduced to a mere three, all assembled in the same direction.

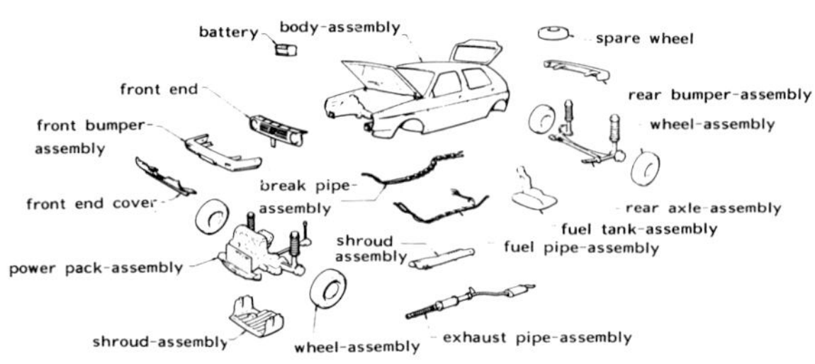

Figure 13
Conception of automatic assembly of the new "rabbit"

Of all conceivable measures, those carried out on individual parts have the least potential for rationalization although it is frequently possible for them to be implemented on existing products with but slight modification effort and expense.

Economical automation of assembly can be greatly assisted by assembly-oriented product design, which is itself supplemented or ultimately made complete by assembly-oriented fastening or joining methods.

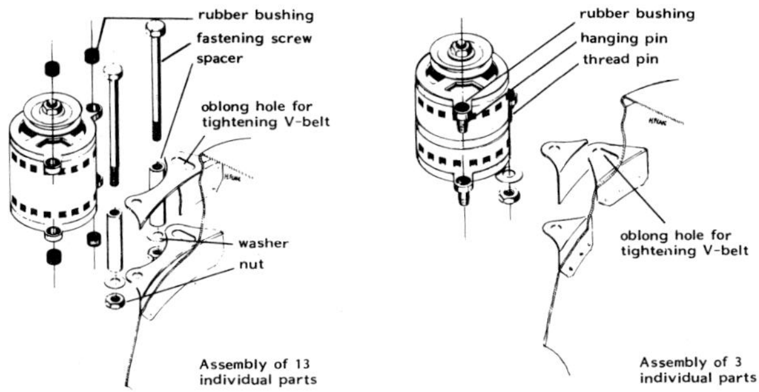

Figure 14
Assembly oriented design applied on the attachement of a motor drum of a washing machine

Modern assembly-oriented joining methods
Forty-one percent of the respondents interviewed in the representative survey mentioned above named simpler joints as a contribution to reducing standard assembly times. This stresses the importance of assembly-oriented joining methods. Established fastening methods such as screws may well answer the requirements on a joint such as reliability, convenient servicing and so on, but fastenings of this type entail high efforts and expense both in making the individual parts of a product and automating its assembly [14].

Other means of substantially simplifying assembly operations are provided by fastening techniques based on recent advances in plastics engineering. Techniques such as the insert and outsert methods of fastening do necessitate extra effort in manufacture of the parts themselves, but this is readily compensated by the simplification of subsequent assembly given a certain volume of production.

Developments in the plastics industry not only result in more convenient shapes and lighter weights of parts; they thus enable design of the products to be assembly-oriented from the start [15].

Together with the rules and guidelines for assembly-oriented product design which are gradually becoming known, the trends in joining indicated here are prerequisites to actually designing assembly-oriented products.

OUTLOOK

According to the forecasts, intensifying competition in the world's marketplaces will lead to a strong increase in the numbers of industrial robots for assembly operations in Germany and elsewhere. The principal target for assembly robots will be applications involving high demands on flexibility. In other words, they will be used to solve assembly problems which up to now have not progressed beyond manual work owing to small production volumes, frequent resetting and excessive complexity.

A marked drop in the prices of industrial robots, a precondition for broad economical employment in assembly, will come in the next few years. Predictions are that they will be halved by 1989 (fig. 4). Ultimately the course of events rests on the ability to match the peripherals of assembly systems to the flexibility of robots. In robotic assembly operations thus far, the flexibility inherent in the equipment has hardly been utilized. Use of sensor systems enables the scope and thus the cost of the peripherals to be reduced. On this point, more extensive conversion of research findings into practical results is needes.

LITERATURE

[1] Warnecke, H.-J.; Schraft, R.D.: Industrieroboter, Krausskopf-Verlag, Mainz 1979.

[2] Walther, J.: Möglichkeiten und Grenzen der Montageautomatisierung. VDI-Z 124 (1982) No. 22.

[3] Bässler, R.; Wolf, E.: Industrieroboter für die Produktion. wt-werkstattstechnik 7/83.

[4] Andreasen, M.M.; Kähler, S.; Lund, T.: Design for Assembly. IFS Ltd. and Springer-Verlag, Berlin, Heidelberg, New York, Tokyo, 1983.

[6] Lund, T.; Kähler, S.: Design for Assembly. Assembly Automation 1983, p. 333-349, Tokyo.

[7] Dilling, H. J.: Methodisches Rationalisieren von Fertigungsprozessen am Beispiel der montagegerechten Produktgestaltung. Dissertation Darmstadt University, 1978.

[8] Schraft, R. D.: Assembly oriented Design-Condition for Successful Automation. Proceedings of 3rd ICAA, p. 155-164, Stuttgart, May 1982.

[9] Schraft, R. D.; Bässler, R.: Possibilities to realize assembly-orientend product design. Proceedings of the 5th ICAA, Paris, May 1984.

[10] Studie zur Untersuchung der Einsatzmöglichkeiten von flexibel automatisierten Montagesystemen in der industriellen Produktion unter Beachtung der technischen, arbeitsorganisatorischen und sozialen Voraussetzungen einer menschengerechten Gestaltung der Arbeit. Government study, prepared by IPA, IAO, GfAH u.a., Bonn, February 1984.

[11] Swift, K. G.: A computer based design consultation system. Assembly Automation, No. 3, p. 151-154, August 1983.

[12] Abele, E.; Walther, J.; Bässler, R.: Montageautomatisierung mit Industrierobotern. Flexible Automation, Mikrocentrum Eindhoven/Enschede, May 1983.

[13] Köhne, F.: Praxis der Montageautomatisierung am Beispiel der Automobilindustrie. Tagung Praxis der Montageautomatisierung, Fellbach. VDI-Berichte 479, p. 1-13, February 1983.

[14] Lotter, B.: Montageerweiterte ABC-Analyse. 5. Deutscher Montagekongress, München, p. 53-71, October 1983.

[15] Kohn, H.: Fertigung von Kunststoffkraftstoffbehältern (KKB) mit hohem Automatisierungsgrad. Kunststoffe im Automobilbau, p. 157-191. VDI-Verlag, Düsseldorf, 1983.

ARTIFICIAL INTELLIGENCE IN MANUFACTURING
T. Bernold (editor)
Elsevier Science Publishers B.V. (North-Holland)

INTEGRATING IMAGE AND NATURAL LANGUAGE UNDERSTANDING

Bernd Neumann

Fachbereich Informatik
Universität Hamburg
Hamburg, W. Germany

As computer vision technology and natural language interfaces are developed for advanced applications in automation and robotics, the need for integrating the two arises. This contribution focusses on the internal representation of the physical environment which is both required for image and language understanding. It may serve as a common database supporting not only vision and language capabilities but also other advanced AI processes such as planning and spatial reasoning. The problem of mapping quantitative geometrical data into qualitative natural language expressions and vice versa is considered in detail.

1. INTRODUCTION

In this paper I discuss problems arising in the boundary area of Computer Vision (CV) and Natural Language (NL). It is well known that both disciplines - CV and NL - are major research areas of Artificial Intelligence (AI), each abundant with difficult and as yet insolved problems. Naturally, the problem of integrating the two areas has not yet received much attention, and it might be argued that it is premature to consider integration problem at all. But I hope to show that the connection of CV and NL is interesting and rewarding both from a scientific and an application-oriented point of view.

Section 2 of the paper will deal with the application aspect. I shall briefly describe current applications of CV and NL in manufacturing and point out possible uses of integrated systems. In section 3 I shall discuss a major conceptual building-block of an integrated system termed "geometrical scene description" (GSD). It provides an internal representation of the physical environment and hence is an important common base for image and natural language understanding. The output of a CV system is essentially in the format of a GSD. In fact, a CV system is generally defined to construct a GSD. Major difficulties arise, however, when connecting a NL system to a GSD, i.e. when talking about scenes. Section 4 addresses this issue and presents a framework for mapping scenes into words and vice versa.

2. VISION AND NATURAL LANGUAGE SYSTEMS IN MANUFACTURING

Before discussing integrated applications of CV and NL in manufacturing, it is useful to review the applications of each individual field, as there is a considerable number of such systems in practical use. From the respective trents we may hope to gather some insight concerning future uses of integrated systems.

COMPUTER VISION APPLICATIONS

CV systems are employed in manufacturing for a steadily increasing range of applications. There are several comprehensive surveys available (NAGEL 85, PAU 84, NEUMANN 84a) with a detailed discussion of the current possibilities and limitations. In this paper I can only address a few issues.

Present-day CV systems are predominantly employed for quality control and visual inspection tasks, and the vast majority of such applications takes place in the automobile industry. The tasks are usually fairly simple compared with the problems which are tackled by experimental systems in research laboratories. For example, for an inspection of - say - an axis assembly, certain checkpoints are brought into camera view at predefined positions and certain image properties are verified in predefined image areas. Such systems basically perform signal processing. They rarely exhibit typical AI features such as explicite object models or hypothesis testing.

There is, however, an increasing number of CV applications in robot sensing which do successfully employ AI technology, in particular for object recognition, automatic assembly and robot navigation. These tasks are typically much more demanding than an inspection task as objects must be recognized irrespective of varying positions in 3D space. Hence a distinction between object features (in 3D) and the corresponding image features (in 2D) becomes important. Also more than a single object may be visible and the need to distinguish many objects and recognize meaningful object configurations arises. Generally speaking, CV systems for advanced applications perform recognition at several levels of abstraction as indicated in Fig. 1. Levels above object recognition are still in an experimental stage. They are important, however, as they indicate an ongoing development and also suggest an interface to other cognitive processes, in particular to a NL system.

NATURAL LANGUAGE APPLICATIONS

NL systems play a role in manufacturing quite different from CV systems. They come into play whenever man has to communicate with machines. As Wahlster points out in a recent survey (WAHLSTER 84), the currently available technology permits useful applications of NL interfaces, restricted to limited domains but featuring impressive cooperative capabilities. Advanced applications include database and expert system interfaces. Three NL systems (deserving that name) are commercially available at the end of 1985.

It appears as if commands, queries or specifications to a machine in a manufacturing environment pose a much simpler problem than any of these advanced tasks. However, the traditional definition of NL understanding does not include speech recognition, i.e. the acoustic part of language understanding. Unfortunately the state of the art of speech recognition does not yet permit casually spoken input to a NL system and hence must be considered the bottle-neck for a wide class of applications where spoken input is essential.

As there is considerable industrial interest in speech recognition devices and several research efforts are underway, one can expect practicable solutions within the next decade. Hence it is foreseeable that also in manufacturing NL interfaces will be widely applicable.

INTEGRATED APPLICATIONS OF VISION AND NATURAL LANGUAGE

Both CV and NL are active research fields and the best commercial systems are still not good enough for many potential applications, hence one cannot expect mature examples of integrated CV and NL systems. In fact, the work which has been

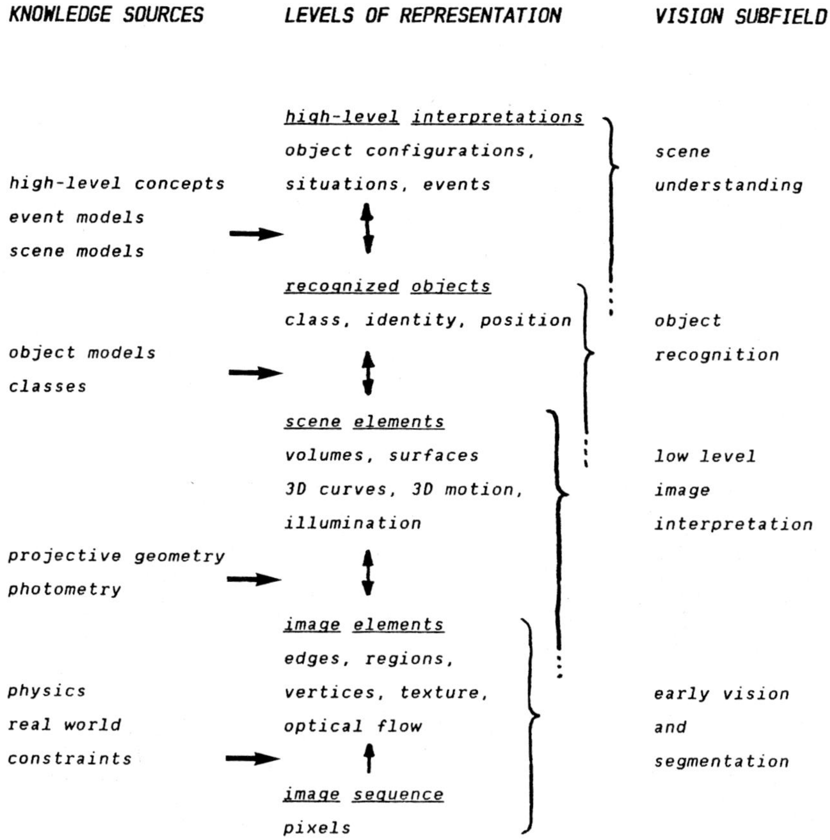

Figure 1: Levels of abstraction in a vision system

done in the boundary area of vision and natural language is largely exploratory and experimental. But nevertheless, it has revealed interesting scientific links and possibilities for integrated applications.

One kind of application is NL communication with a CV system. This is an important task as the output of advanced vision systems may be too complex to be rendered by lists or graphics or other adhoc means. It is also an obvious application given existing CV and NL systems. Example systems of this kind are described in SCHEFE and PRETSCHNER 81, NEUMANN 82 and BAJCSY et al. 85. A NL interface to a CV system does not make much sense, if the visual domain is highly artificial, e.g. in xray quality control. It may be very useful, however, for sufficiently natural domains such as a factory work space or an automatic assembly station. A "say-what-you-see" system could report the presence and location of interesting objects or events, dangerous situations, system problems, etc.

Another kind of integration comes about by using a NL interface and visual sensors as components in a common system framework, say for process control. Without elaborating concrete examples, it can be seen that language and vision may not be installed as independent components. Whenever NL and CV (sub-)systems have a common domain and deal with the same conceptual entities (e.g. work pieces), they must also share internal representations and have means to communicate. This point will be discussed further in the next sections.

Robot systems provide probably the best examples for integrated NL and CV applications. In fact, robotics is that field of AI which involves virtually all other AI disciplines. Robotics has been defined as "the intelligent connection of perception and action" (BRADY 84). Brady's "mechanic's mate" (BRADY and AGRE 84) is an intelligent robot in this sense. It is a proposed manipulator system with vision and language understanding as well as planning and reasoning capabilities. While such product is not around the corner, it clearly marks a long-term goal and gives some indication of the problems which must be handled by an integrated system. Consider a situation where a human worker asks the robot to "pick up that wrench on the table". The command must be interpreted by the NL system of the robot in terms of an action involving an object to be recognized by the CV system. Furthermore, object recognition is controlled by the qualitative geometrical constraints of the NL expression "on the table" as well as other constraints implied by the situational context.

As a first step towards a "mechanic's mate" the teach-in process of present-day robots could be improved by providing integrated NL and CV control. For example, one could advise the robot to "move the gripper above the housing" instead of manually placing the gripper into the desired position. Clearly not all robot tasks can be conveniently specified verbally and assisted by CV. But there certainly are applications where this approach would constitute an improvement over "teaching-by-showing" or keyboard methods which prevail today.

3. REPRESENTING THE PHYSICAL ENVIRONMENT

Physical objects are the subject matter common to image and language understanding and hence provide an obvious link for integrating the two subsystems. Of course, NL may also deal with many other, possibly abstract subjects - beliefs, intentions, explanations, etc. But it is the internal representation of the physical environment, to be exact, of the visual properties of the environment, which connects CV and NL. This representation has been termed "geometrical scene description" (GSD) in NEUMANN and NOVAK 83 because of the prevalence of quantitative geometrical information. A GSD consists essentially of

- the (time-varying) locations of physical surfaces, and
- classified or identified objects.

In other words: A GSD describes the scene geometry and names recognized objects. The geometrical specifications are quantitative as opposed to symbolic. They can be given in terms of the spatio-temporal coordinates of CAD/CAM-type object shape descriptions, for example in terms of triangulated surfaces. This characteristic of a GSD is important as NL expressions about the same matter tend to be qualitative, hence do not have an explicite representation in a GSD. The problem of mapping qualitative NL predicates into quantitative data and vice versa is discussed in the next section.

A GSD is the natural output of a vision system in the narrow sense. It contains the information provided at the level of recognized objects (see Fig. 1), but it does not necessarily contain higher-level interpretations. A GSD can be considered a canonical way of representing vision system output as it conforms to the definition of a vision system commonly used in this field: "Vision is to know what is where by looking" (MARR 81).

Considering again the "mechanic's mate" or for this matter any advanced model of cognitive activities, we can see that a GSD may also play an important role for other integrated systems. This is illustrated in Fig. 2, where several cognitive processes are shown to interact with a GSD, providing additional support for the claim that a GSD is a key structure for integration.

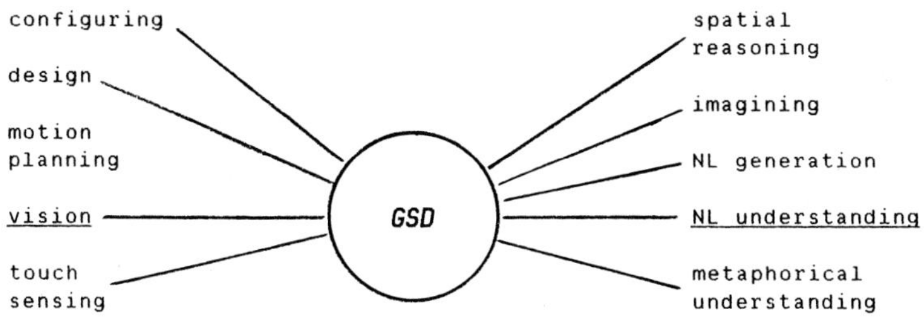

Figure 2: The geometrical scene description as the basis for cognitive activities

4. TALKING ABOUT SCENES

It has been pointed out that a CV system generates a quantitative geometric scene description as a natural output of image analysis processes up to the level of object recognition. Unfortunately, this is not the format of the internal representations traditonally used for generating or understanding NL text about the same subject matter. Most NL systems which have been developed so far fullfill their purpose with rather crude representations of scene geometry (if any). Typically, geometrical information is represented in terms of symbolic qualitative relations (e.g. "left-of" or "near") which stand for the corresponding NL expressions. The NL systems HAM-RPM (V.HAHN et al. 80) and HAM-ANS (HOEPPNER et al. 83) are notable exceptions as they employ a scene coordinate system from which qualitative relations are computed. I shall now present a generalized approach to this problem which has been developed for NAOS, a system for describing time-varying scenes (NEUMANN 84b).

VERBALIZING SCENE GEOMETRY

We consider the problem of verbalizing a time-varying geometrical scene description (GSD), in particular the mapping of quantitative scene data into qualitative predicates corresponding to NL expressions. For example, we would like to obtain a descriptive sentence "Robby approached the box left of the table" from a corresponding GSD.

Motion verbs (e.g. "approach") and spatial prepositions (e.g. "left-of") are important linguistic means for describing the spatio-temporal properties of a scene. They are examples of a structured system of qualitative predicates with varying degrees of specialization. It is interesting, both by conceptual and implementation-related reasons, that all such predicates are based on a limited set of primitives. One can think of the primitives as the set which results from applying a limited number of basic qualitative predicates to the observable quantities of a scene.

The observable quantities are:

- position, distance (w.r.t. a reference position)
- orientation, angle (w.r.t. a reference angle)
- temporal derivatives of the above

The basic qualitative predicates are:

- constancy
- range
- greater/less

Thus, we have primitives such as:

- constant location ("stand still","move")
- derivative of distance less than zero ("approach", "slow-down")
- location in behind-range ("behind")

From primitives one can construct composite predicates as, for example, "bounce". This motion verb (which is one of the examples in the early work of BADLER 75) can be composed of a "downward-motion" followed by an "upward-motion", both of which are primitive.

NL has several other means which help to translate spatial or temporal scene properties into sentences. For example, modifiers or linguistic hedges may be used for fine-tuning, as for example in "very close" or "he approached rapidly". Also there are ways to specify the temporal relationship of events by temporal connectives ("before", "after"), see MARBURGER et al. 81 for further details.

It is important to note that the qualitative predicates discussed above constitute only necessary conditions for using certain NL expressions. In general, verbalization is governed by many rules and constraints depending on semantical aspects other than geometry, and also, of course, on syntax and pragmatics. We have emphasized the problem of obtaining qualitative predicates from qualitative scene data as this is a key problem which has to be solved for integrating image and NL understanding.

THE INVERSE: QUANTITATIVE INTERPRETATION OF NATURAL LANGUAGE

We now turn to the inverse problem: How can quantitative geometrical data be obtained from NL expressions? This may be quite important for an integrated NL and CV system, as some of the applications mentioned in section 2 show. If we want the CV system to retrieve the location of - say - "the wrench on the table", there must be a way to map the qualitative NL expression "on the table" into a range of potential wrench locations. Similarly, if we want the robot to place a wrench "on the table", it is necessary to map the NL expression into quantitative data for robot motion control.

The examples show two aspects of the problem. First, as a qualitative predicate is typically many-to-one, the inverse is one-to-many, i.e. gives a range of possible values which in most cases may be conveniently represented by constraints. Hence further processing, in particular the top-down control of the CV system, must be constraint-based. This calls for a special system design as realized, for example, in the ACRONYM system (BROOKS 81).

The second aspect concerns tasks which require a unique geometrical interpretation of a NL expression in spite of the one-to-many mapping. Robot motion control is one such task, visualization for spatial reasoning is another. In order to be able to choose a particular value from many possible values, additional information has to be brought to bear. This can be done by providing knowledge about prototypical situations, i.e. about quantitative values which are typically associated with a NL expression given a certain situation. For example, the center-coordinate of the table may be associated with the expression "on the table" given an empty table-top. Prototypical knowledge may also be useful for guiding the search of a CV system within the boundaries of NL constraints.

The problem of quantitative NL interpretation is an open research topic with few contributions so far. But there is a large body of literature about the closely related visualization problem, notably from psychologists and linguists (see BLOCK 81 for a representative collection).

5. CONCLUSIONS

I have shown that several advanced AI applications, particularly in robotics, call for the integration of image and NL understanding. The conceptual framework of the integrated system is centered around a representation of the scene geometry as a common referential database for both image analysis and language understanding. While such database is a natural interface to vision, it poses some problems on the NL side which have not yet received much attention so far. One problem is the specification of word meaning in terms of quantitative scene geometry. I have presented a unifying approach which grew out of the work on NAOS, a system for NL description of time-varying scenes (NEUMANN 84b). The key idea is to define basic qualitative predicates on the time-varying scene geometry, from which a large class of word meanings can be derived. The inverse path leading from a NL expression to a quantitative scene description is not yet well developed. Additional knowledge about typical scenes may be required to resolve the inherent ambiguity of one-to-many mappings.

6. REFERENCES

Badler 75
Temporal Scene Analysis: Conceptual Descriptions of Object Movements
N.I. Badler
Report TR 80, Department of Computer Science University of Toronto, Toronto/Canada 1975

Bajcsy et al. 85
LandScan: A Natural language and Computer Vision System for Analyzing Aerial Images
R. Bajcsy, A. Joshi, E. Krotkov, A. Zwarico
IJCAI-85, 1985, 919-921

Block 81
Imagery
N. Block (ed.)
The MIT Press, Cambridge/Mass. 1981

Brady 84
Artificial Intelligence and Robotics
M. Brady
Artificial Intelligence, 1984

Brady and Agre 84
The Mechanic's Mate
M. Brady, P.E. Agre
ECAI-84: Advances in Artificial Intelligence, Elsevier 1984, 681-696

Brooks 81
Symbolic Reasoning Among 3-D Models and 2-D Images
R.A.Brooks
in: J.M. Brady (ed.), Computer Vision, North-Holland Publishing Co. Amsterdam, 1981, 285-348

Hoeppner et al. 83
Beyond Domain Independence: Experience with the Development of a German Language Access System to Highly Diverse Background Systems
W. Hoeppner, T. Christaller, H. Marburger, K. Morik, B. Nebel, M. O'Leary,

W. Wahlster
IJCAI-83, 1983, 588-594
Marburger et al. 81
Natural Language Inquiries about Motion in an Automatically Analyzed Traffic Scene
H. Marburger, B. Neumann, H.-J. Novak
in: J. Siekmann (ed.), GWAI-81, Informatik-Fachberichte 47, Springer Verlag Berlin-Heidelberg-New York, 1981, 79-87
Marr 81
Vision
D. Marr
Freeman, 1981
Nagel 85
Wissensgestuetzte Ansaetze beim maschinellen Sehen: Helfen sie in der Praxis?
H.-H. Nagel
in: W. Brauer, B. Radig (Hrsg.), Wissensbasierte Systeme, Informatik Fachberichte 112, Springer 1985, 170-198
Neumann 82
Towards Natural Language Description of Real-World Image Sequences
B. Neumann
GI - 12. Jahrestagung, Informatik Fachberichte 57, Springer Berlin/Heidelberg/New York 1982, 349-358
Neumann 84a
Vision Systems: State of the Art and Prospects
Bernd Neumann
IfI-HH-M-124, FB Informatik, Universitaet Hamburg, 1984
Neumann 84b
Natural Language Description of Time-Varying Scenes
B. Neumann
in: D. Waltz (ed.), Advances in Natural Language Processes, Volume I, 1984 (to appear), also as FBI-HH-B-105/84, Fachbereich Informatik, Universitaet Hamburg, 1984
Neumann and Novak 83
Event Models for Recognition and Natural Language Description of Events in Real-world Image Sequences
B. Neumann, H.-J. Novak
IJCAI-83, 724-726
Pau 84
Approaches to Industrial Image Processing and Their Limitations
L.F. Pau
Electronics and Power, February 1984
Schefe and Pretschner 81
SWYSS - A Natural Language Question-Answering System for Scene Analysis
P.Schefe, B.Pretschner
GWAI-81, Informatik Fachberichte 47, Springer 81, 69-78
Wahlster 84
Cooperative Access Systems
W. Wahlster
Future Generation Computer Systems, Vol. 1 No. 2, 1984, 103-111
v.Hahn et al. 80
The Anatomy of the Natural Language Dialogue System HAM-RPM
W.v.Hahn, W.Hoeppner, A.Jameson, W.Wahlster
in: L.Bolc (ed.) Natural Language Based Computer Systems, Muenchen, Hanser/McMillan 1980, 119-253

ARTIFICIAL INTELLIGENCE IN MANUFACTURING
T. Bernold (editor)
Elsevier Science Publishers B.V. (North-Holland)

TACTILE WITH VISION: AN INTEGRATED APPROACH TO ROBOT SENSING

N. Ghani

MARI Advanced Microelectronics Limited
32 Grainger Park Road
Newcastle upon Tyne NE4 8RY
UK

This case study discusses the use of a combination of tactile and visual imaging in the sensory control of a robot applied to workpiece handling in FMS. A vision system gives an initial assessment of workpiece position and orientation, sufficient to allow an attempt by the robot to grasp the object. During the grasping process, the tactile sensor confirms, refines or modifies the initial assessment. The problems of integrating these two types of input are discussed.

1 INTRODUCTION

At the present time, the term AI tends to be synonymous with "expert systems", primarily because this appears to be one area that currently is being exploited commercially. However in the context of manufacturing applications, the robotics and sensory perception aspects of AI have been the earliest points of attention.

At MIT, the definition of an intelligent robot is one that "intelligently connects perception to action" [1]. This case study is concerned with the development of a robot system capable of intelligent action within the limited context of a specific class of manufacturing problem, based on perception of a well-defined environment.

The use of vision techniques for robot perception has attracted much research and development effort over the years [2]. Indeed, robots with some degree of visual sensing capability are now available commercially. On the other hand, the control, integration and programming problems associated with using such systems still limit their widespread application.

The development of additional or alternative sensors to widen the perception capability of robots has attracted much interest. For example, we now see robots with force and torque sensors built in.

Tactile sensors which provide an additional imaging technique are relatively less developed [3]. Most of the effort in this area has gone into the sensing techniques involved - how to acquire the tactile image which is basically a stress map of the contact surfaces between the gripper and the object. The problems of interpreting this type of image in the context of robot control are even less understood.

This case study is based on a project being carried out under the Computer Integrated Manufacturing (CIM) subprogramme of ESPRIT. Project 278 which started in January 1985 addresses the problems of putting together a robot system using both vision and tactile senses in its control scheme. The project, wishing to address real manufacturing problems, set as one of its objectives the task of constructing such a robot system which would be applied to the seemingly simple task of acquiring and orienting randomly presented workpieces.

2 THE MANUFACTURING REQUIREMENT

Most manufacturing processes require the presentation of correctly oriented workpieces. Unfortunately it is rare for a workpiece to arrive at a workplace in a known orientation and positioned exactly. When production volume is high, specially engineered solutions can be justified, for example for preserving orientation through workpiece-specific carriers, or for restoring orientation through specialised devices such as vibratory bowl feeders, or a combination of both.

With the trend towards flexible manufacturing, implying frequent changes of workpiece, the specially engineered solution is less viable. Equally the alternative of using manual labour for orientation and feeding is not attractive in this context [4].

There is therefore a need for a means of acquiring unoriented workpieces from a general-purpose carrier (pallet, belt, bin) in a manner that is flexible, reliable, easily adaptable to new workpieces, and capable of cycling fast enough. This is the prime industrial need which is addressed in this case study. The research project which is presented here is in the process of developing a sensor-based robot system to meet this need.

3 THE SENSING PROBLEM

The basic problem is to determine the position and orientation of a (known) workpiece with sufficient accuracy to allow proper grasping, transfer and placement. This aparently simple problem can easily be complicated in a number of ways:

- more than one type of workpiece possible at any time
- more than one workpiece in the field of operation.
- touching components
- overlapping objects
- objects not lying on a horizontal plane
- objects with no stable states

The worst case is obviously that of a bin holding a variety of jumbled parts [5]. To provide a more realistic target for a three year timescale, the project concentrates on the simpler aspects of the problem. It is considered however that dealing with these complicating factors in practice may require significant advances in the application of knowledge-based techniques to the problem.

In the simple case then we have a known object on its own, lying on a horizontal plane, in one of a number of known stable states. Taking a pragmatic approach, the problem then reduces to one of determining (a) which stable state the object is in, (b) the rotational angle of that stable state in the horizontal plane from some fixed axis, and (c) the position on the horizontal plane of a point on the object. From these, and knowledge of the object, it is possible to compute the complete positional and orientation data for the robot to grasp the object.

Using a vision system with a camera mounted above the plane we can begin to determine these data. The problems that arise here include:

- calibration
- optical distortion
- sensitivity to lighting
- image skew
- object rotation
- ambiguity

These problems may be tackled by various techniques [6] but in the final analysis, they tend to introduce some degree of uncertainty into the output of a vision system.

Take for example a rectangular valve block lying on its side. It may be impossible to determine from a view from above which of the two possible orientations and which two possible stable states are correct. The position of the block as computed by the vision system may have a degree of error associated with it because of some of the above-mentioned problems.

The role of a tactile sensor becomes clearer here. Incorporated into the gripping surfaces of the robot's end effector, a pair of tactile sensor pads should provide further information about the object during the process of grasping. (The gripper is assumed to be a two-finger device with parallel motion.) Assuming that distinguishing features are present on the gripping surfaces, the tactile image data should help resolve orientation ambiguity. Furthermore, as the gripper position is known, and the tactile data is derived from direct contact between object and gripper, the accuracy of any positional information that can be determined from the tactile data is limited only by the resolution of the tactile sensor pad, and the accuracy of the robot's positional sensing mechanisms.

Thus, while the vision system provides a 2-D view of the entire object though with some uncertainty, the tactile sensor should provide a 2-D contact image of subsections of the objects' surface, with well-defined accuracy.

The problem then is how to integrate these two complementary types of data (as well as other sensory inputs) into the robot control scheme such that the system can be adapted to a new part with a minimum of time/cost/effort/specialist skills, and can carry out the task reliably, with as low a cycle time as possible.

In the context of reliability, it should be noted that a subsidiary function of the tactile sensors is to detect movement (i.e. slippage) of the object during transfer.

4 THE INTEGRATION PROBLEM

One of the features which (in my opinion) distinguishes AI from traditional approaches is that with AI, it is not necessary to have

all the necessary input data before attempting to compute the results - there is usually some means of acquiring more relevant data during the computation (i.e. "asking intelligent questions").

Another "AI-type" feature is the ability to backtrack, where the response to further data may be to cause previous decisions to be altered.

In studying the relationship between vision and tactile sensing, both these features are apparent. In order to acquire tactile images it is necessary to have sufficient knowledge of the object's whereabouts, sufficient to allow an attempt to touch it. This leads naturally to a strategy based on using vision to provide this initial knowledge.

Using this knowledge, the system can then approach the object with the gripper and its tactile sensors, and by touching it, attempt to answer the question "did I see right?". Ideally, the tactile image should be acquired at the desired gripping point, so that if the answer is "yes", the object can then be picked up with a minimum of delay. However, it is quite possible that the ideal gripping position for an object may be a completely featureless surface, which would provide no further information (other than possibly confirming the presence and attitude of that surface). It is also possible that the features which are present are too symmetrical to provide useful orientation information.

In order not to restrict workpiece spectrum, we need to logically separate the tasks of tactile image acquisition and of workpiece grasping. The possibility of combining the two should be regarded as an optimisation.

The crucial issue here is how the question "did I see right?" should actually be asked in practice. It is particularly important in the context of cycle time, as it has to be answered in real time, and the robot cannot proceed until it is answered.

The obvious form is "given what I thought I saw, am I feeling what I should expect to feel?". This of course assumes that it is possible to associate, with each stable state of the object, one or more tactile image templates, together with whatever information is needed to move the gripper to each position where each template would be expected to apply. The setting up of all this data in the system's knowledge base is of course non-trivial and is discussed later in Section 5.

Thus in an ideal situation, the robot would move to the object according to vision data, and attempt to grip it. The tactile image pair obtained in the grip attempt would conform exactly to the template (i.e. the features correspond, and position within sensor window is exact) and the robot would then proceed with the transfer.

In moving away from the ideal, the first possibility to consider is a situation where the features match, but the images are offset in some way from the templates. This implies that the vision data was qualitatively correct, but quantitatively inaccurate. The obvious response here is to correct for the offset, either before picking the object up, or during transfer and placement.

Note that in the situation where the tactile image is acquired in a different place, any such offset can be corrected in moving to the grip point.

The other situation is that there is no match at all with the templates. There are two possible reasons for this. The obvious one is that the initial choice of stable state made from vision data was incorrect, requiring the system to backtrack. The other is that the choice was correct, but because of quantitative errors the plane of the gripper surfaces did not match the object surfaces correctly. (The latter could probably be minimized through the use of compliance in the robot arm/gripper design).

Another issue which arises here is the problem of deciding, when, during the process of closing the gripper fingers, should the tactile images actually be sampled, and accepted as valid. Sampling should clearly take place when the gripping force reaches some pre-defined limit, and the images have stabilised by then. At this stage, a simple test such as gripper finger separation could be applied to check if the images should be regarded as valid.

Note that in the above, the vision system is assumed to have done its job after the initial view and analysis. One potential problem with backtracking is that the object might have been displaced during acquisition of a tactile image. This would require the backtracking to go even further, in that the vision system would need to take another view and start again. Furthermore, the gripper and arm would need to be moved out of the field of view. With its obvious implication on cycle times, this is a situation to be avoided. With careful force thresholding during grasp it may be possible to keep any displacement to within the same order of magnitude as the quantitative errors from the vision system.

5 OTHER IMPLEMENTATION ISSUES

The previous section concentrated on the strategies envisaged for integrating tactile sensing into an overall scheme for accurately sensing the position of and attitude of a workpiece. Here a number of more practical issues are examined, which highlight the fact that there is still a long way to go.

5.1 Viable Tactile Sensors

There is clearly a need for low-cost, rugged, compact, reliable and easily manufactured tactile sensors which provide sufficient resolution, if such a system is to be accepted in a manufacturing environment. At present a few sensors are available commercially, but with limited resolution and high cost. ESPRIT project 278 is also looking at the sensor technology. Optical techniques appear to offer the most promise, especially in the future when the technology of integrated optoelectronics becomes better developed.

At present it is not clear if binary sensors are sufficient for the task, or whether some gray-level capability is needed. It is felt that binary images may be all that are needed for tactile image image analysis, but that a gray-level sensor with low-level adaptive thresholding may offer better image acquisition. Another factor relevant to this is the amount of crosstalk between adjacent points on the sensor array (which depends on the sensing technique used).

For example, a sensor with high crosstalk is likely to need adaptive thresholding to acquire the best image.

5.2 Tactile Image Analysis

Whilst the analysis and interpretation of tactile images may appear to be simpler than for vision (given smaller images, no shadows, no occlusion, and no depth problem), the fact that each tactile image is a restricted window view of the object's surface gives rise to problems peculiar to tactile.

Take for example a reference image acquired during teaching which includes a feature represented as a closed contour. It is quite possible that during operation the window boundary cuts across the feature, resulting in a working image containing a contour segment which then has to be matched with the original closed contour. Of course, the reverse is also possible. The matching algorithm (and the subsequent computation of window offset) has to be signifcantly more complex. There is also the possibility that the set of features seen during operation may not map directly into the set seen during teaching. This may well require the use of a model-based approach, with the reference features derived from a geometric description in a CAD data base, instead of acquiring the reference by teaching.

5.3 Teaching/Programming

For such a system to be cost-effective in the small batch situation, the cost of adapting to each new workpiece must be minimised. This implies a system that can be programmed or taught easily and quickly, with minimum downtime.

The ideal situation would be to carry out off-line programming on a CAD system, with task-oriented programming tools and associated graphics and simulation capabilities. This would also require the ability to derive, from a CAD model of each workpiece, sufficient data in the right form needed by both the vision and the tactile subsystems. This is still some way away.

Current commercially available technology provides for robot-level action oriented languages, combined with manual-guided teaching. In the context of sensor-based systems, such languages and teaching methods have to be extended to handle sensor programming (and the more sophisticated data handling and decision making that will be needed). As sensor programming currently involves extensive teaching (to build up sufficient data for the statistical pattern recognition techniques typically used) [7], it is clear that the extended methods must handle teaching and programming of robot and sensors in an integrated manner.

A number of pragmatic approaches suggest themselves:

- the robot itself could be used to automate and speed up sensor teaching. Once taught to grasp an object in a particular stable state, the robot could systematically move the object on the working plane to present a variety of views to the vision system. At the same time it could acquire samples of tactile image pairs needed by the tactile system.

- a certain amount of adaptive learning may be possible. As the tactile system should, during operation, provide feedback as to the success or failure of the vision analysis, it should be possible to use operational data to enhance the initial image statistics. With this approach, the initial teaching could be reduced, with an initially high re-try rate improving with time.

5.4 Error Recovery

In order to be acceptable to industry, the system must be able to cope with the majority of error situations without needing human intervention. A number of typical error situations can be envisaged.

(a) Slippage - on the assumption that this can be detected by monitoring changes in the tactile images during transfer, there are two obvious reactions. The gripping force could be increased, although clearly there must be an upper limit to avoid damage. The other response is to correct the trajectory taking into account the offset due to slippage. A more difficult situation might arise if the object has slipped to a position where the gripper cannot continue to grasp it reliably. Detection and correction of this is not straightforward.

(b) dropped object: - (the worst case of slippage). This is easy to detect, but corrective action may be rather more difficult to specify as a general solution.

(c) collision - this is part of the more general problem of collision avoidance and trajectory planning which are significant research areas in themselves.

(d) dead-end - it is quite possible that in spite of backtracking, there may be situations where the tactile input cannot be reconciled with the vision data. This might arise out of shortcomings in the image interpretation algorithms. It may be necessary to backtrack even further, e.g. use an alternative camera, change the lighting.

5.5 Workpiece Spectrum

A practical system should be able to handle any shape or form of workpiece (within defined limits of size and weight) without any need for specialised programming skills on the part of the user. The primary requirement here is for image analysis algorithms able to perform matching and measurement without having to be told about special features. Clearly, in the case of the tactile system, it may still be necessary for the user to choose the best points on an object for the tactile sensor to work on; this should however be part of the teaching or production planning process.

5.6 Unconstrained Bin-Picking

Most of the discussion in this paper is based on the assumption that parts are presented separately on a horizontal plane. In practice, this is less of a constraint than it would appear: it is a lot easier to maintain separation than it is to preserve orientation. Removal of this constraint, even partially, (e.g. touching but non-overlapping parts) complicates matters. Apart from the need for more sophisticated vision algorithms even for initial assessment, the need for knowledge-based planning becomes apparent for carrying out the task of grasping a part. There is now the question of graspability, and a need for alternative grasp positions or even a spectrum of grasp positions. Strategies may be needed for re-grasping an object which may have had to be grasped initially in an inappropriate manner to move it away from a neighbouring part.

6 CONCLUSIONS

The fact that this case study highlights problems and approaches to solutions rather than definite solutions, reflects the early stage of the research work that it is based on. Clearly it is too early to draw any conclusions other than the obvious one that the development of a sensor-based robot system able to act intelligently, even in a limited problem domain such as workpiece acquisition and orientation, still has a long way to go. There appears to be good scope for exploitation of tactile sensing, but there are still some major problems to be solved before it can be generally applied and used.

7 ACKNOWLEDGEMENTS

The project mentioned in this paper is partly funded by the European Community under the ESPRIT Programme. The interest, help and encouragment given by Mrs. P. MacConaill and Dr. A. Ikonomoupoulos of the ESPRIT Task Force have been invaluable.

The views expressed in this paper are primarily those of the author, and are not necessarily shared by all members of the Consortium involved in the project. On the other hand, I have benefitted from discussions with Bob Campbell and Zep Rzepczynski of MARI, and with Hans Lindner of IPA (who were instrumental in defining the manufacturing problem domain to be addressed in the project).

8 REFERENCES

[1] Brady, J.M., Intelligent robots: connecting perception to action, in: Winston, P.H. and Prendergast, K.A. (eds), The AI Business (MIT Press, Cambridge, 1984).

[2] Hollinghum, J., Machine Vision (IFS Pubs, UK, 1984).

[3] Harmon, L.D., Automated tactile sensing, Int. Jnl. Robotics Research 1/2 (1982) 3-32.

[4] Warnecke, H.J., Development trends in automation: benefits to small and medium sized manufacturing enterprises, Ann. CIRP 29/2 (1980) 455-467.

[5] Birk, J.R. et al, An orienting robot for feeding workpieces stored in bins, IEEE Trans SMC 11/2 (1981) 151-160.

[6] Raymond, C., Donath, M. and Olson, W.R., Problems of vision directed robots in an unstructured parts handling environment, Proc 13th Int. Symp. Ind. Robots 2 (1983) 14/1-14/18.

[7] Pot, J., Coiffer, P. and Rives, P., Comparison of five methods for the recognition of industrial parts, in Rooks, B., Developments in Robotics (1983).

EXPERT SYSTEMS

ARTIFICIAL INTELLIGENCE IN MANUFACTURING
T. Bernold (editor)
Elsevier Science Publishers B.V. (North-Holland)

ASSESSMENT OF THE APPLICATION OF EXPERT SYSTEMS IN FLEXIBLE MANUFACTURING

Gisela Kiratli

Werkzeugmaschinenlabor (WZL)
Technical University of Aachen
F.R.G.

The following report will discuss the applicability and impacts of knowledge-based systems in flexible manufacturing using the Integrated Manufacturing and Assembly System at the Machine Tool Laboratory (WZL), Technical University of Aachen, as a potential application case.

INTRODUCTION

This paper does not describe an already existing expert system. But the impacts of these new software technologies, based on AI methods and influencing flexible manufacturing are discussed in the following. At first the aspects of automation will be pointed out before describing a realized flexible manufacturing system at the WZL at Aachen. This system will be the basis of the discussion. At least the applicability of expert systems will be proved using the following examples:

- a three level expert system for controlling a Flexible Manufacturing System (FMS) and

- an expert system for diagnosis in FMS.

AUTOMATION IN MANUFACTURING

The worldwide increasing pressure of competition in mechanical engineering requires economical manufacturing, a high degree of quality and a short time of delivery. Additionally the brief cycles of innovation together with smaller batch sizes forced the companies to introduce flexible, computer aided automation in manufacturing.

Having a look at the development of automation (Fig. 1) there is to be seen that in DNC-systems (Direct Numerical Control) the paper tape, used by the former NC-machines, had been replaced. The main task of DNC-systems is the central computerized management and the transmission of NC-programs to the single numerical controls. The task of FMS is automation of material flow and information processing in generally. This leads to future concepts of computer integrated manufacturing, which imply an integration of disposition, organization and production, and the linking of information flow between the different factory areas. In the future computer aided methods will influence all levels in manufacturing.

One important precondition for the spreading of FMS is the advanced standardization and modularization of the necessary system components. The mechanical components of such a system are offered in form of building blocks, whereas there are no standard control systems for FMS. In that subject software means a bottleneck for further progress - like it does everywhere in industrial data processing. So for every FMS concept there has to be programmed a new control system. Software has to be developed by conventional programming methods, broadening the gap between software and hardware development costs.

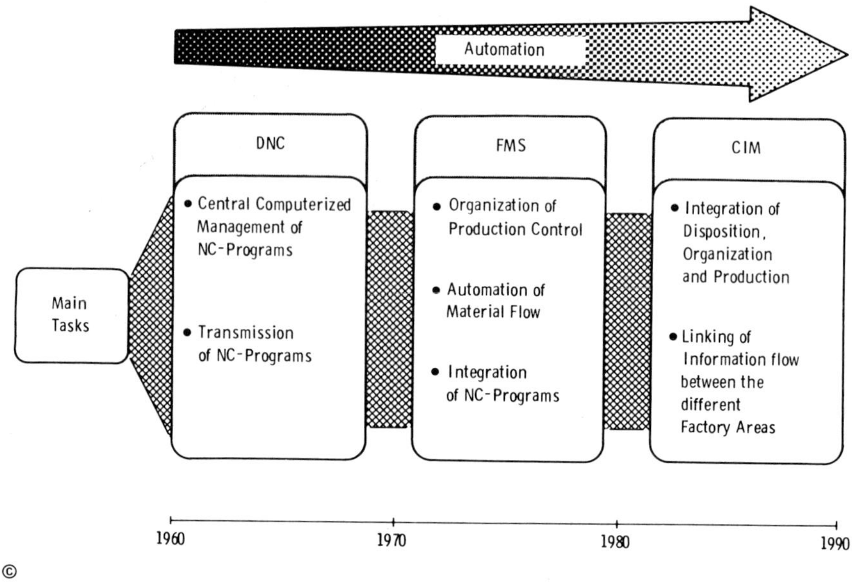

Figure 1: Development of computer integrated manufacturing

MODULARIZATION OF FMS

Although different in concept and realization the hardware and software components of a FMS are to be splitted up into always similar main groups. Fig. 2 shows the hardware components of a FMS. It consists of several work stations (e. g. manufacturing machines, measuring stations, washing stations) which are linked together with buffer and storage systems by means of automatic material flow (e. g. workpiece, tool, fixture flow, etc.). Depending on the flexibility of the system, different workpieces can be processed at a time, whereby these workpieces can pass through the system by different paths according to the machining requirements of the respective workpiece. A joint control system coordinates, supervises and controls all the material flow and machining processes during automatic machining.

A hierarchical concept consisting of several control levels (Fig. 3) has been generally approved as a control structure of a FMS. The disposition level is responsible for production planning and control, that means the rough capacity planning. The computers on the supervision and system level perform job scheduling of the orders, the machine occupancy planning and the organizational process control.

The single mechanical components of the FMS are equipped with largely intelligent, autonomous control devices which perform a large part of the functions for automatic machining and automatic material flow. The supervisory manufacturing computer coordinates and activates automatically the processes in the system and enables the operator to supervise and intervent out of a central level.

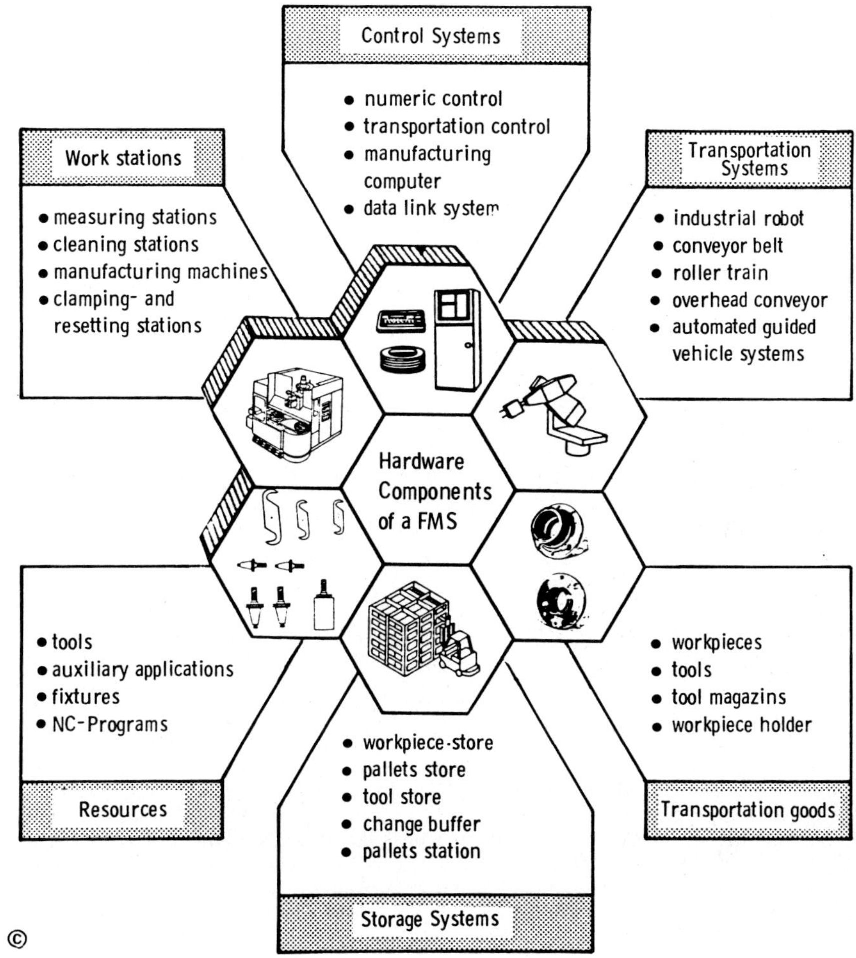

Figure 2: Hardware components of a FMS

Fig. 4 shows the tasks and the functions of the control system software in FMS. These are dispositive, organizational and operative control tasks, the monitoring and administrative functions and at least the communicative tasks concerning the main fraim, the operator and the device control. The main aspect of FMS means automation of material and information flow whose connections can be described by specific features. These features are concerned with the special system layout, the transportation system, the machining task and the properties of the device controls.

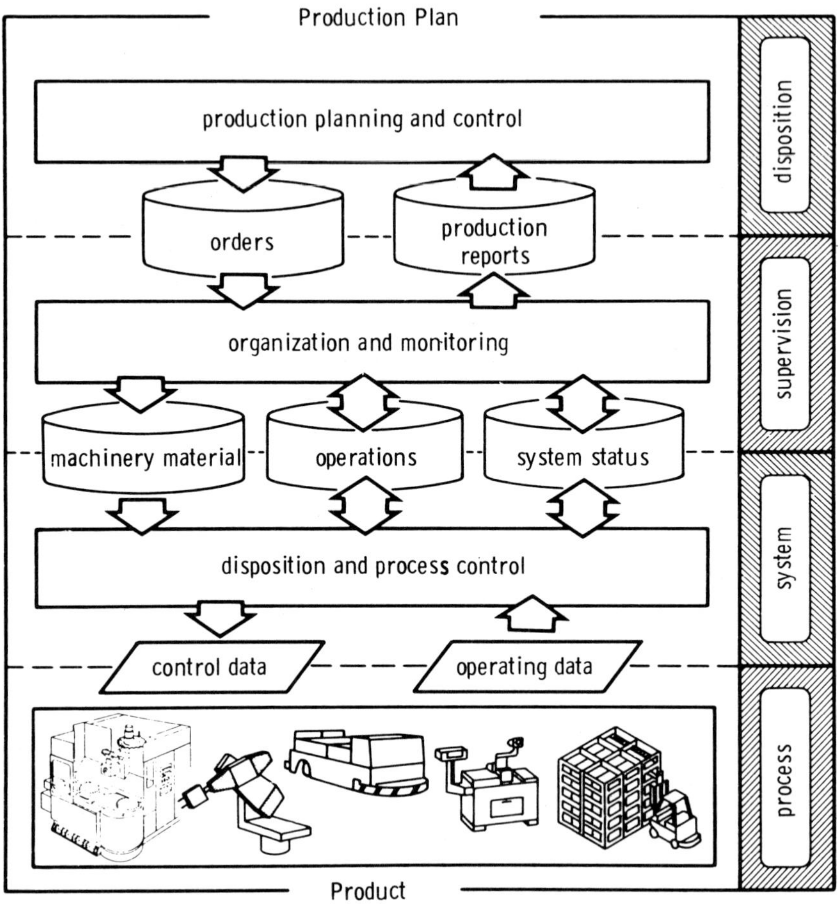

Figure 3: Hierarchically structured control of a manufacturing system

UNIVERSAL CONTROL SOFTWARE

In this regard there has been developed a universal control software at the Machine Tool Laboratory at Aachen which can be adapted to different FMS-configura-

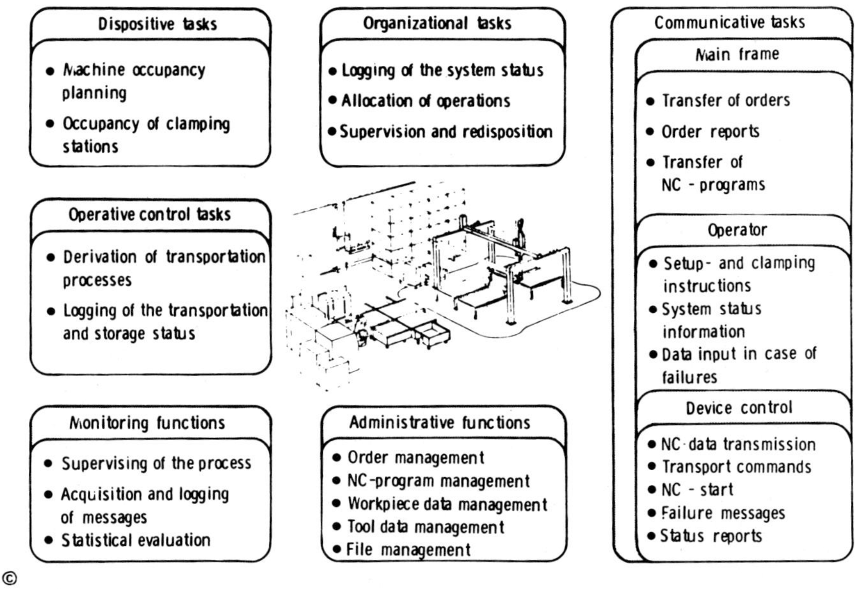

Figure 4: Functions of the control system software

tions by parametrization. This software has been used for the first time to control an Integrated Flexible Manufacturing and Assembly System (Fig. 5).

The main components of this system are the processing centre and the assembly station. The system store contains blanks, semi-finished and finished parts. A floor transport cart loads and unloads the different stations.

The universal control software is a modular software package existing of different software function blocks (Fig. 6). These tasks are processing independently of each other and exchange informations in a common area of memory. The informations are distinguished in invariable ones describing the system and variable ones describing the actual system states. One task during generating the control software deals with defining the material and information flow according to system-layout. The necessary conceptional phase, which fixes system-layout and transport ways, is happening by using paper and pencil. In the existing solution there is no computer aided method. The result of this first step, e. g. type and number of work stations, type and number of transportation units, will be programmed with computer aided methods, by dialogue. After complete description of the machining system, the global parameters and development of new software parts (e. g. the adaption of failure reports) the generation phase will begin.

These parts from the new control software for a special FMS after they had been compiled and linked together.

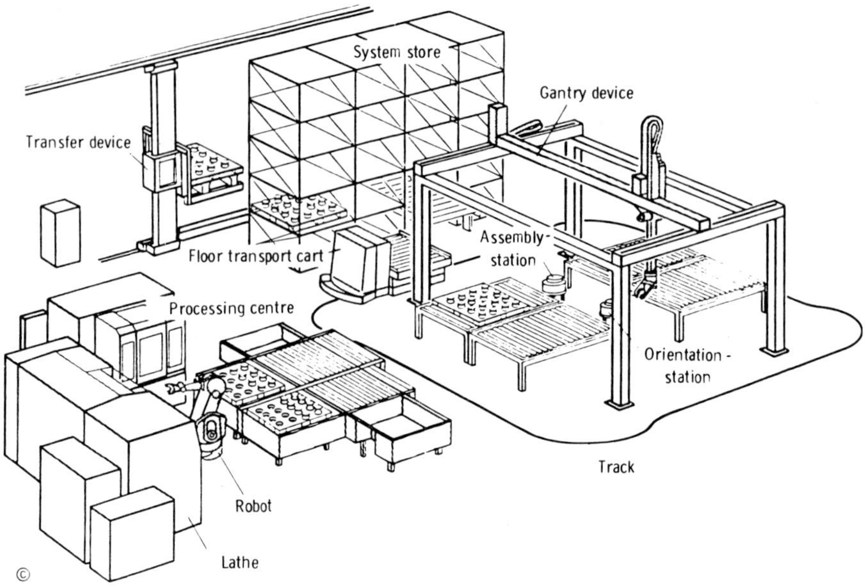

Figure 5: Integrated Manufacturing and Assembly System

The development of this briefly described control software has not been finished. The system serves as a prototype which does not consider all controlling aspects (e. g. tool planning).

Experiences during development and application of this universal control system emphasized the difficulties and problems of developing software with conventional programming methods. Especially during the planning phase this matter became evident. Again and again new conditions and applications had to be considered which influenced already finished development parts, doing the fatiguing job of editing, compiling, linking and testing. The program job has been done by using the high level programming language PASCAL.

Because of the universal character of the control software the generation phase is done by using a lot of branches in the program. That means strong interconnections and dependencies between single program parts. Therefore adding further alternatives, e. g. a new tpye of transportation unit, requires high efforts because the dependencies of former development steps have to be considered, and that is not an easy job. Especially in conventional programming methods there is a lack in testing tools concerning the examination of a specific class of properties. That means e. g., when adding a new alternative dealing with transportation to be listed up for recognizing contradictions and conflicts. Conventional programming methods do not include the examination of the pre-history of an event. Therefore it is very hard to check all the consequences of additional conditions.

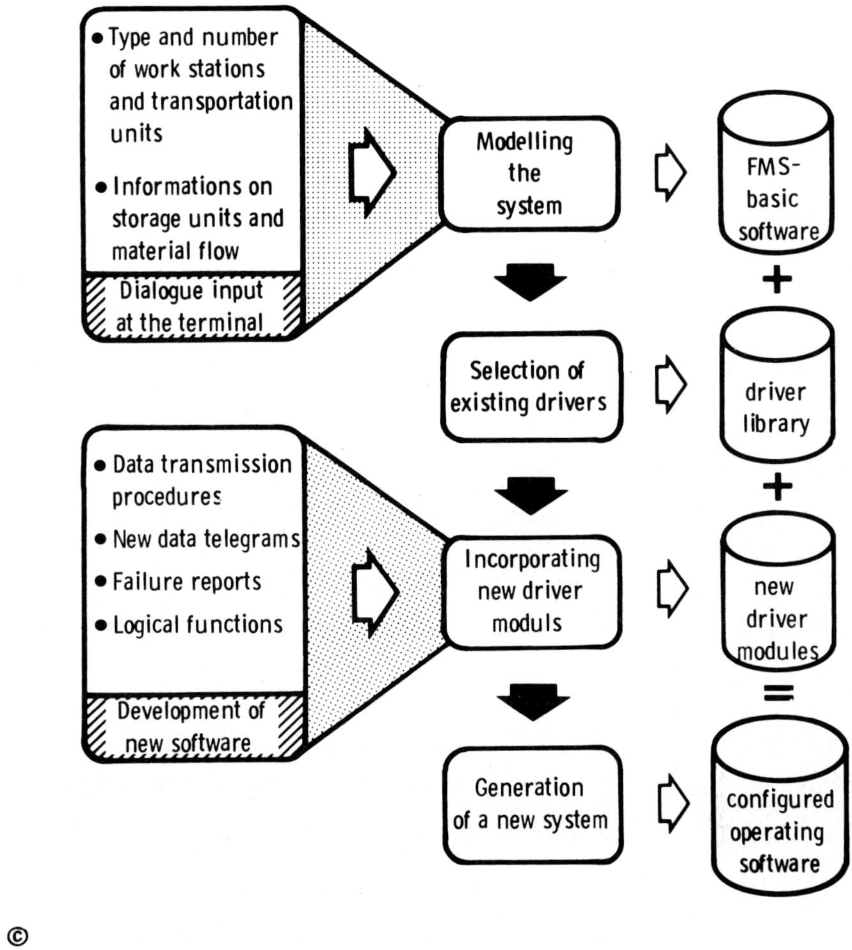

Figure 6: Configuration of the operating software

AN EXPERT SYSTEM FOR THE CONTROL SYSTEM

As a result of all these deficiencies grew the whish to find an aid which is able to solve the enumerated disadvantages and problems. This could be a knowledge-based system, a special area in Artificial Intelligence. Using this systems the sequential programming method will be replaced by a procedure, which distinguishes between execution, that means the inference machine, and the problem task, described in the knowledge base. The special world of application is splitted up into single structure elements which seem to be without relation and which are to be combined together during a specific operation. Changing the knowledge base does not imply the time-wasting mechanism as it does the algorithm method.

Today there are offered shell systems for the development of expert systems. These shells contain tools which allow - to a certain degree - examining the plausibility of later added knowledge parts in comparison with the existing ones.

But there are some preconditions for the application of expert systems in manufacturing. E. g. it must be possible to embed existing conventional software (like driver programs for controlling the single units). Additionally it is necessary to develop a suitable interface between the device controls (on machining level) and the expert system.

A further advantage (Fig. 7) during construction of an expert system is the possibility of partly modelling and simulating the problem after a short time (method of rapid prototyping). This lightens up the burden of planning and developing the software for a universal FMS control software. Especially the implementation of the explanation tool - a special feature of knowledge-based systems - enables to approve the plausibility of partial solutions.

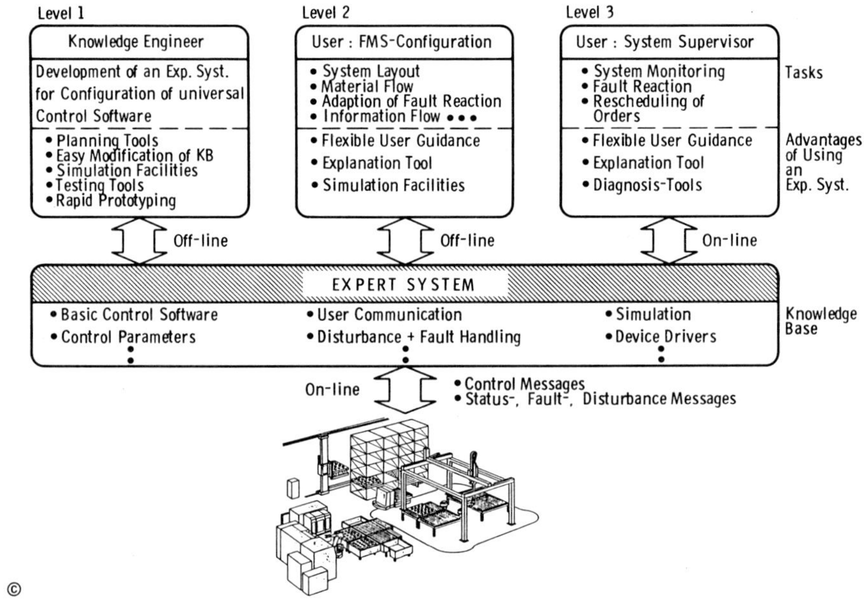

Figure 7: User Levels of expert system application for FMS-control

The next user level of this expert system supports the configuration of a special FMS-application. Here the user can profit by the facilities of such a system during the process of configuration. The reasons for this are the possibility of flexible user guidance on the one hand and the explanation tool on the other hand, which e. g. can be asked for the necessity of configuration steps. This method is able to replace planning with paper and pencil. Graphical supply can figure e. g. the specific material and information flows of the selected system layout. Modelling events in these flows enables examining the generation result in an early state. Means for optimizing are to be applied during configuration process.

The third user level is the one of the system supervisor who serves the specific control software configurated in the former generation process. His task is to supervise the manufacturing system and to react to extraordinary events, like disturbance - and fault-messages (e. g. by order redisposition). The expert system supplies him with advice contributing to find the cause of disturbance. The advice can be took of experiences resulting out of processing of the FMS or out of former applications. Additionally there is to be attached a valence for every reaction in accordance to the value of its evidence. Flexible user-guidance and the explanation tool allows a more effective proceeding on this user level, too.

EXPERT SYSTEM AS A TOOL FOR DIAGNOSIS

The FMS-control software, developed with conventional programming methods, as described before, must not react in realtime. Time-critical tasks are solved on the device level by the single device controls. The supervision of processing now is made by figuring a stilisized copy of the system events on a coloured graphic terminal. Thus there can be described processing events and status informations, as well as disturbances and fault messages. But the supervisor has to react on these messages by himself, he has to plan and perform measures independently. That means, with the aid of graphic disturbance events are only made visible, but the exploration for the very reasons are to be obtained by the user.

For this reason there is another application case for expert systems, that means in the section of diagnosis and reaction to disturbance messages. In this subject diagnosis means evaluation of fault messages and the derivation of reasons. The processing of a lot of informations is necessary for this object. Especially it is very significant that a disturbance message often has not only a single reason. Thus a definit diagnosis is possible just after asking for further conditions. Statistical frequency of faults and heuristical values of experience are to be considered.

Naturally it is not possible to execute these requirements with the aid of conventional programming methods, especially regarding that during the processing of an FMS there can happen individual fault reasons. So a diagnosis system must be flexible concerning the addibility of further experience values.

Problems of this form cannot be programmed procedurally because the complexity of diagnosis in FMS exceeds the degree of solvability with conventional programming methods. The combinatorical explosion regarding all the possible fault messages and their reasons require another proceeding.

These requirements can be obtained by applying an expert system (Fig. 8), whose logical structure gives the possibility to describe problems not in form of a sequential algorithm, but in relational and functional dependencies. Thus the way of finding out a solution is another one. That is the way of reasoning with the aid of existing facts to further ones, which can be verified under this consideration. The selected way of solution can be different at every new session.

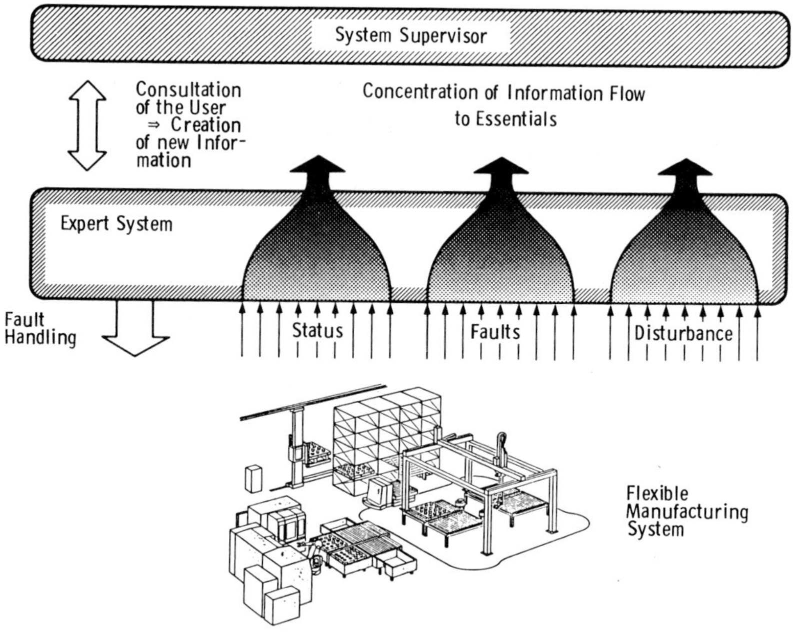

Figure 8: Using an expert system as diagnosis tool in flexible manufacturing

Additionally, there is the possibility to use alternative ways and - by consulting the user - to choose this one, which leads to the very reason of fault.

A further step of automation is that the expert system derivates a treatment for a special fault or disturbance message. It can perform these reactions by itself or propose it to the user. This proceeding requires some preconditions. On the one hand there is the necessity of a realistic timing of the expert system, because it must react in a time as short as possible to avoid a system break down. On the other hand the progress in sensor techniques plays an important role. The higher the degree of automation the more the cognitive facilities of human being have to be assumed by suitable sensors. These sensors give to the expert system a figure of the controlling system. Thus this capability of sensors is a direct measure for the quality of an expert system for diagnosis.

PROSPECT

In the previous explanations there are discussed the possible applications of expert systems on the supervisory and system level of a FMS. But there is to be imagined its usage on the upper level, which obtains the disposition and the production control. Especially informations resulting out of the lower levels can be prepared and made available for administrative decisions of the management level, that means, further management functions can be derived with the aid of a knowledge-based system. Only the relevant informations are filtered out of the production process. An explanation tool will lighten up single events in this case, too.

These coarsely figured conceptions lead to the idea of connected expert systems whose knowledge bases partly overlap each other and which are suited for the requirements of the respective user circle. Thus there exists the possibility of integrating further administration tasks, e. g. the tool planning. But there are many problems to be considered, especially the necessity of parallel and real-time processing, the problem of maintenance of large and frequently used knowledge bases and the multi-user capabilities.

SUMMARY

Expert systems are a suitable instrument for a consequent conception of FMS. On the one hand the user has supports in fulfilling tasks he had to obtain without tools in former times, on the other hand the computer performs tasks, which could hardly done by conventional software systems.

Especially knowledge-based systems for FMS are adequate means for joining to CIM and smooth the way to the "Factory of the Future", provided that there will happen further progress on the field of AI-software, AI-hardware and the field of sensor techniques.

REFERENCES

/1/ Kohen, E.: Aspekte einer rechnergeführten Fertigung. VDI-Bildungswerk, 1984.

/2/ Weck, M.; Kohen E.: Konfigurierbare Steuerungssoftware für flexible Fertigungssysteme. Industrie-Anzeiger 106 (1984) Nr. 83, S. 25 - 30, Girardet-Verlag Essen.

/3/ Weck, M. et. alteri: Flexible Fertigungsanlagen. Report during "Aachener Werkzeugmaschinen-Kolloquium" (June 1984), Technical University of Aachen.

ARTIFICIAL INTELLIGENCE IN MANUFACTURING
T. Bernold (editor)
Elsevier Science Publishers B.V. (North-Holland)

The Development Of ALADIN
An Expert System For Aluminum Alloy Design

Martha L. Farinacci [**], Mark S. Fox [*], Ingemar Hulthage [*], Michael D. Rychener [*]

** Process Control and Computer Tech. Div., Alcoa Laboratories, Pgh., Pa., USA
* Intelligent Systems Laboratory, Robotics Inst., Carnegie-Mellon Univ., Pgh., Pa. USA

ALADIN is an expert system that aids metallurgists in the design of new aluminum alloys. The system is a hybrid of several artificial intelligence techniques. Declarative representations are used for metallurgical concepts. The design procedure is encoded in a rule-base, with the potential for user participation. The system is based on the hypothesize-and-test model, and a constraint-based search is used to find alternative designs. Strategies are included for resolving the interactions among conflicting subproblems. Metallurgical relationships cover a broad range of symbolic and numerical types of reasoning. In this article, an overview of the project is given, including discussions of knowledge acquisition techniques, design decisions and implementation.

1. Introduction

ALADIN (ALuminum Alloy Design INventor) is an expert system that aids metallurgists in the design of new aluminum alloys. The system can be operated in several modes. As a decision support system, it will accept alloy property targets as input and suggest alloying additives, processing methods or microstructural features to meet the targets. As a design assistant, it can evaluate designs supplied by a metallurgist, or provide information that is useful for design from a knowledge bank.

ALADIN was developed through the combined efforts of three organizations: The Intelligent Systems Lab of the Robotics Institute of Carnegie-Mellon University, the Artificial Intelligence group at Alcoa Laboratories, and the Alloy Technology Division of Alcoa Laboratories.

Design and development of the system began in January, 1984 and continues at the time of this writting. While it is difficult to evaluate a system that is under development, it is hoped that ALADIN will provide many important benefits when completed. The amount of knowledge required to successfully develop new materials is so great that individuals often must supplement their private knowledge with information from books, journals and specialized consultants. ALADIN, when used as a knowledge bank, will provide another source of valued information. There is some hope that by fusing together multiple sources of knowledge from different experts, a system will be developed that exceeds the capabilities of individual experts. The development of new alloys now requires testing many alternatives before one adequate solution is found. An attempt is being made to develop a more complete understanding of the behavior of alloys through the development of quantitative models. These models will , in turn, enable development to proceed with fewer tests. ALADIN will serve as a collection point for these models and also the older empirical methods. Hence ALADIN will aid in the development of new materials at a reduced cost and time.

In this report, a case study of the ALADIN project to date is given.

2. Assessment of Artificial Intelligence for Alloy Design

Computers have long been used to solve engineering and science problems in the metals industries. For example, finite element models of numerous manufacturing processes, including rolling, extrusion, forming, forging, casting and smelting, have been developed [9]. With these models, the effects of processing on the shape of the material, internal stresses, temperatures, and currents (in the case of smelting) are predicted. Simulation methods are usually used to study processing on a larger scale. Like finite element methods, these models are predictive although they often deal with multiple processing steps. They can determine the effects of material handling practices, plant layout, equipment design and operating practices on product characteristics and equipment performance. Another major application of computing in the metals industry is process or numerical control. Tool adjustments are made in order to control the shape and quality of the product being produced. Often the control models are based, in part, on the equations underlying the predictive models mentioned earlier [2].

The classical applications of computing just described share several important features:

- The problems are primarily numerical.
- The processes are well understood quantitatively, and equations are available to describe effects.
- Well defined algorithms are available to solve the equations and make the appropriate predictions.

Engineering problems with these qualities are usually solved with models implemented in a procedural language such as FORTRAN.

Alloy design, in contrast to these problems, is characterized by the following features:

- Quantitative models for calculating the properties of a proposed design are often not available, or insufficient data prevent their use.
- Models that are available contain a high degree of uncertainty.
- The design process, as practiced by people, often involves reasoning about images appearing under a microscope, abstract concepts, interpretations and heuristic rules of thumb based on experience. The reasoning is highly symbolic as opposed to numeric.
- The alloy design process, as practiced by people, appears to be similar to the diagnostic problems discussed in artificial intelligence literature. When given a new set of alloy specifications, designers identify an existing alloy, determine in what way the existing alloy fails to meet specifications, and try to find a cause for the discrepancy. Alloy design problems are

more complex than diagnostic problems, however, since a great deal of reasoning is required to construct a solution even after the cause of the problem is identified.

- No fixed algorithm seems to be used in design. Instead, metallurgists seem to search through the space of design variables with guidance from heuristic rules and models. The flow of reasoning depends on what knowledge is available and how reliable that knowledge is at the time the design problem is specified.

It was because of these features that an artificial intelligence approach was used in the alloy design problem.

Although alloy design has a few features that make it a natural artificial intelligence application, there was some early evidence of difficulty. The solution of a single problem takes weeks or often months. This is because the reasoning is complex, many options must be explored, and vital missing information must be searched for. Because of this, it was recognized that knowledge acquisition would be difficult. Furthermore, design involves reasoning about time. Processing of the metal involves changes to the structure of the material which influence later production steps. In other words, the time and sequence in which changes are made is important. It is known that planning problems of this sort are difficult to solve. However, it was felt that the benefits of an alloy design system would outweigh the risks involved in its development.

3. Alloy Design Reasoning

Before beginning a technical discussion of the ALADIN system, an overview of the reasoning used in alloy design will be given. This overview does not require any knowledge of metallurgy. The introduction of a few basic scientific concepts will make later descriptions of the system clearer.

The alloy designer usually begins his problem with a set of constraints on the physical properties of the material he is to make. The constraints are often one-sided, as in: strength must be higher than x and density must be lower than y. Sometimes the constraints involve properties that are not numerical such as machinability or surface appearance. Specifications usually involve many properties and correspond to a material that does not yet exist. This is because the design problem comes from a customer or the corporation who would like something that exceeds the capabilities of everything on the market. The design problem may be overconstrained - there may be no material that can meet the specifications given. It is difficult to know in advance which problems are impossible to solve. The designer is expected to do the best he can and come as close as possible to meeting the targets.

Once property constraints are specified, designers usually select some baseline alloy to begin their search for a solution. Design will then involve finding alterations to the baseline that change the characteristics of the material in the direction of the targets. There are several different strategies that people use for selecting a baseline. Some look for a commercial alloy that comes as close as possible to meeting the targets, or the experimental alloy from the most recent iteration in the current design problem. Still others begin with a simple, commercially pure aluminum alloy and design from basic principles.

After the baseline is selected, designers look for changes that can be made to the material in order to improve the properties. The things that can be changed directly are:

- COMPOSITION - the elements added and the amounts

- PROCESS - the fabrication steps used, their sequence, and such specifications as method, temperature or time

There are rules that indicate the effects of these design choices on properties. Some examples are:

- *IF* an element with low atomic number is added *THEN* density will decrease

- *IF* Mg is added *THEN* strength will increase

However, the most powerful rules involve reasoning about the microstructure of the alloy. Some examples of these rules are:

- *IF* The aging process is done for a long time *THEN* equilibrium precipitates will form

- *IF* equilibrium precipitates are present *THEN* they are usually incoherent

- *IF* incoherent precipitates are present *THEN* they may form on the grain boundary

- *IF* precipitates are on the grain boundary and strength is medium or high *THEN* elongation and fracture-toughness are low

Metallurgists often illustrate the knowledge they work with in a graph (figure [3-1]). Nodes correspond to the four classes of knowledge that characterize an alloy: the composition, processing, structure and properties. Arrows indicate the relationships that are used so heavily in design. The rules just mentioned are examples of these relationships. For instance, the first rule: "*IF* an element with low atomic number is added *THEN* density will decrease", corresponds to the arrow from composition to property.

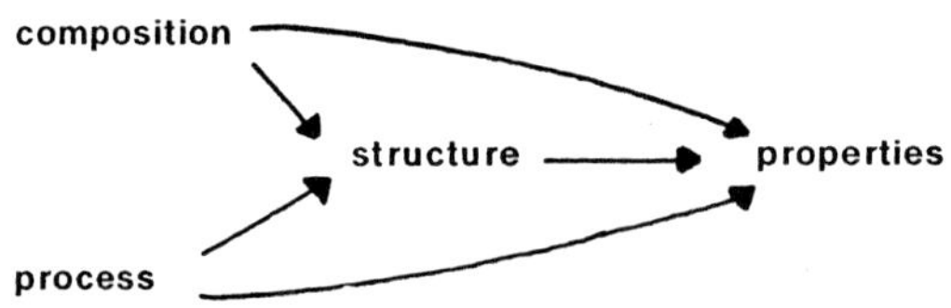

Figure 3-1: alloy design graph

Designers seem to apply rules in an opportunistic fashion. Whenever rules are identified that will make some progress in solving the problem, those rules are applied. There are a few patterns to the search process, however. Rules that involve reasoning about structure are given a preference over rules that do not. Also, some property targets are viewed as being more important than others, and rules that deal with important targets are used first.

Because metallurgists are asked to develop materials with properties outside of the range of existing alloys, they must reason about designs which have never been tested and whose behavior is not known with certainty. Designers try to fill this gap in existing knowledge with general models, speculation, extrapolation of known trends, and analogy with existing materials. Even after applying these methods, however, it is usually impossible to determine exactly what composition and processing specification will meet the targets. The designer identifies a family of alloys that is likely to meet the targets, and makes these alloys in the laboratory. Sometimes, one member in the family will have the desired characteristics. Often, all tests will fail to meet targets exactly, but analysis of the data results in more accurate rules that can form the basis for a better set of experiments in the next iteration.

4. Case Study of ALADIN

ALADIN was designed and developed using a rapid prototype approach. During the first year, 1984, metallurgists were interviewed, the alloy design methods were characterized, a system was designed and the program written. At the end of 1984, a demonstration of the system was given to the metallurgists and an assessment of the project was made. This assessment involved technical design decisions as well as staffing levels and the approach used for knowledge acquisition. Several changes were made to the design and to the organization of the project during 1985. The design changes are now being incorporated into the software, and it is expected that the second iteration will be completed by the end of 1985.

4.1. First Iteration

During the first iteration, many alloy designers were involved in the knowledge acquisition process. During the first two months, five people were interviewed who had done work on a variety of different alloy systems. An attempt was made to find the common approaches used by all designers. By focusing on these, it was hoped that the system could be developed in a general way and could handle many different design strategies. After the initial interviews, a single prototype problem was identified and short term goals were established to develop a system based on the prototype. The problem selected was the first iteration of aluminum lithium alloys. These alloys are now under development at Alcoa and are targeted for future aerospace applications. This choice was made for several technical and nontechnical reasons. The aluminum lithium alloys are important to Alcoa, so management support was expected to be high. Furthermore, many experts were available and anxious to help on the project. The reasoning used on the aluminum lithium project covered a broad range of empirical and theoretical ideas, so a system based on the prototype could be extended to other problems easily. Finally, since the company had only recently begun research on this alloy system, it was speculated that there were many design alternatives, not yet explored by human designers, that the ALADIN system could study. After the prototype problem was selected, a few more experts were interviewed, and a team of five metallurgists was finally selected for more detailed knowledge acquisition.

During the summer and autumn of 1984, the first iteration ALADIN system was designed. Because of tight deadlines for the implementation, several simplifications were made.

The solution method was modelled as a variant of generate & test [7]. This involved iteration of three steps in a fixed sequence:

- Hypothesis-Generation - find actions to move the alloy in the direction of targets
- Hypothesis-Selection - select the best option
- Hypothesis-Evaluation - predict the properties of the alloy and identify deviations from target

Metallurgy rules defining the relationships between comosition, processing, structure and properties provided the operators for generation and evaluation. In cases where operators were not available, regression analysis was used to identify useful trends from the alloy data base [8]. The available generation operators often dealt with property targets independently. So, the design problem was decomposed into subproblems - one for each target. The effects of subproblem interaction were not addressed, although it was known that properties were interdependent. The search space was limited to composition choices. Few microstructure and no process rules were built into the first system. Because of this decision, the system contained almost no advanced scientific knowledge. But the developers were able to test and evaluate the use of various problem solving paradigms on this problem before spending too much time learning difficult metallurgical concepts. A hierarchical planning approach was used: an abstract plan containing the types of elements to add was developed before the details of additive percents were derived.

A representation was developed for a database of alloy information and other metallurgical knowledge. The representation was based on the concept of frames. The database included known alloys with information about alloying and impurity elements, physical properties, product forms and applications. Alloy families and series were also stored and linked to the individual alloys. Inheritance across links was used for default reasoning. The alloy database was used to determine standard practices and to identify useful relationships between composition and properties. Representations were also developed for important metallurgical tables and diagrams. For example, phase diagrams indicate the types of equilibrium phases that are present in an alloy for a given composition and temperature. These diagrams are used extensively by alloy designers when selecting alloying additions and processing conditions. Regions of the diagram and important characteristics of the associated phases were represented in the database. Relations between regions were defined and corresponded to transformations between phases.

Two types of data elements were created to hold information about the alloy design problem - the constraint and the hypothesis. Property targets specified by the user were represented as constraints. Although in real problems, early decisions constrain choices that are available later in the search, no attempt was made to represent these dynamic constraints in the first iteration. A representation was also developed for what was called a hypothesis tree. All design choices were posted on this tree, which was developed in a depth first manner. This hypothesis tree was slightly different from the standard search graphs used for artificial intelligence problem solvers [10]. Separate trees were developed for the abstract and the detailed plans. Links were created between nodes in the two trees to indicate dependencies between the two levels of decisions.

Three types of control elements were created for the system - contexts, goals and tasks. A context corresponded to a major phase of the design problem [1]. Six contexts were identified for the first iteration system - problem-definition, search-setup, hypothesis-generation, hypothesis-selection, hypothesis-evaluation and search-termination. Contexts could have a status of active or suspended, and they were activated in a fairly rigid control sequence, as indicated in the flow chart in figure [4-1]. All metallurgy rules checked for the presence of some active context as the first precondition. Hence, the context was used to decompose the system into major subsystems. Goals were used to represent actions to be performed or already attempted [1]. Goals could be arranged into fairly general and/or/all trees and were built dynamically as the domain rules identified required subtasks. Sequence restrictions could be imposed on subgoals that shared the same parent, although these specifications were not required. Several status values were allowed for each goal, including posted (an action that should be performed in the future),

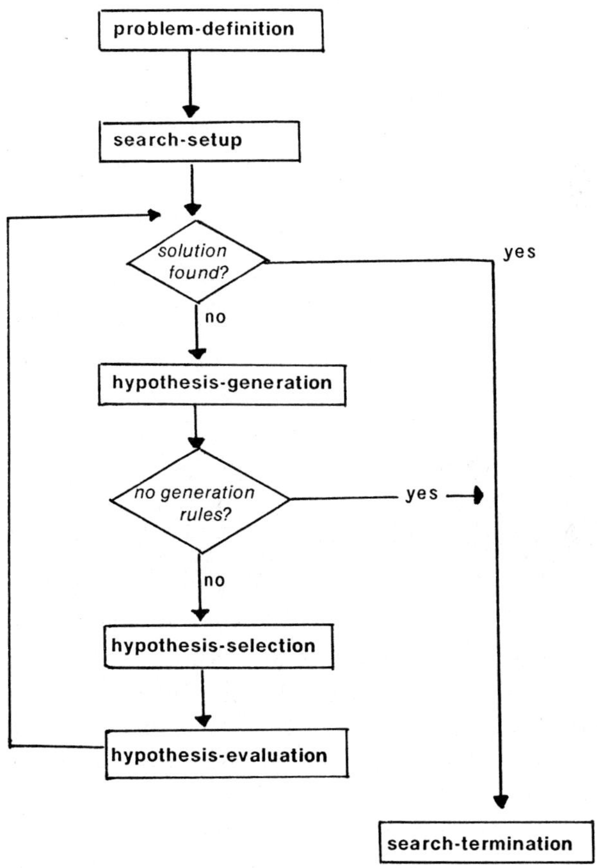

Figure 4-1: first iteration flow of control

active (an action that should be performed now), success (an action that was completed successfully) or failure (an action that was attempted but not completed). All metallurgy rules checked for the presence of an active goal as the second precondition, so the goals allowed for a finer decomposition of the rule set into knowledge sources. A general set of goal management rules was written to update status values throughout the tree whenever the status of a leaf goal node was changed by the domain rules. Finally, task elements were used to represent simple actions to be performed. Unlike goals, tasks could not be linked to form a complex tree structure.

Several conventions and standards were established for the use of control and data elements. Some examples were:

- at any time, only one context and goal should be active
- hypotheses should have a non-nil level, indicating the level of detail in hierarchical planning

Of course, when developing a system, it is always difficult to guarantee that conventions are followed, and failure to follow the practices can lead to subtle bugs that are hard to find. To make the debugging task easier, a set of diagnostic checks was formulated. These rules checked for violations from the standards and reported the errors to the developers.

During the last few months of 1984, the first iteration ALADIN system was implemented. Three languages were integrated to form the system. OPS5 [6], a forward chaining production system, was used for the overall search control, the representations of goals, hypotheses, and constraints and the metallurgy rules. CRL [4], a frame-based knowledge representation system was used to store long term declarative knowledge about alloys, their composition and properties, and also metallurgical tables and diagrams. FRANZ LISP [5] was used for the interface between OPS5 and CRL, for the user interface and also for some numerical procedures. The system was implemented on a Vax 750 running under the Unix operating system.

During the first few months of 1985, an assessment of the ALADIN project was made. Most of the earlier decisions were still found to be correct ones. For example, after a year of knowledge acquisition, it was felt that an artificial intelligence-based approach was a reasonable one to use on the alloy design problem. Similarly, the choice of languages was still felt to be a good one. Even some of the design decisions were still felt to be appropriate. The goal management system, the decomposition of the rules base with contexts and goals, the generate and test approach, the idea of using some problem decomposition according to targets, and the use of hierarchical planning were retained for the second iteration. However, several problems were also identified. The fixed sequence of context activations was unable to respond to problems such as lack of information or failure of goals. The system had to be extended to deal with decisions about structure and process. The first iteration representation of the hypothesis tree was unable to record all of the important details about structure and process decisions. The representation of constraints also had to be generalized to deal with restrictions on composition, process and structure decisions that could be proposed, evaluated and then withdrawn during the problem solving process.

Finally, some problems were identified in the organization of the team and the approach used in knowledge acquisition. Gathering information from five metallurgists was extremely confusing and it was difficult to organize the knowledge in a coherent fashion. The primary method used to teach metallurgists about ALADIN system capabilities was to run demonstrations of the program. However, this proved to be confusing as well since at any time, most planned capabilities were not implemented and also, the user interface was not well developed.

4.2. Second Iteration

The second iteration of the ALADIN system began after the assessment phase. A conversion was made to a newer version of the CRL language and Common LISP [11], running on a VAX VMS system. Then all steps were repeated - knowledge acquisition, system design and implementation. However, the approach was changed based on the first year's experiences.

Major changes were made to the organization and staffing of the project. The aluminum lithium prototype problem was replaced by an even more specific problem that was called the training case. This case consisted of four experiments that were identified and tested during the first iteration of the aluminum lithium project. The ALADIN project goal for 1985 was to characterize and model the reasoning that was used to select those four experiments. The alloy design staff was reduced to two people, who were each allowed more time to spend on the project and given more responsibilities.

One expert had a great deal of experience with the aluminum lithium training case. He was given the responsibility of helping to prepare a description of the reasoning used on the problem. One computer scientist assisted him with this assignment. The metallurgist tried to describe the sequence in which ideas were considered and rejected and to characterize the rules that were used to make the decisions. The computer scientist tried to reformulate this reasoning in terms of the ALADIN design concepts. In other words, the flow of metallurgical ideas and the justifications for decisions were classified as hypothesis evaluation, hypothesis generation using search operators, backtracking, etc. New contexts and goals were identified during this process, and strategies for context switching were formulated. A written report was prepared in which the reasoning for the training case was described in detail. The report was partitioned into sections - approximately half of the sections contained descriptions using metallurgical language; the other half contained formulations of the same ideas using computing concepts. During each meeting, the report was reviewed and corrections were made before work continued. After six iterations, the report was reasonably complete.

Another expert was chosen for his experience in working with computers and for his research interest in developing general models of metallurgical structure/property relationships. His main responsibilities were to help design the knowledge representation system for microstructure and processing and to help incorporate his models into the system. Another computer scientist worked closely with him to identify ways that the models could be smoothly integrated into the existing system. Written reports were again used to ensure that ideas had been understood correctly. This expert was also asked to use the system and to comment on its functionality and accuracy. Because he had more experience in using computers than many of his colleagues, it was expected that this expert would be more tolerant of the simple user interface.

The design of the system was changed in four major ways. A meta space was designed that provided a more rational sequencing of contexts and goals. The hypothesis representation was generalized to deal with decisions about processing and microstructure. A least-commitment decision-making strategy, using multidimensional constraints, was incorporated into the system. Finally, several provisions were made to deal with interactions among subproblems.

The meta space was responsible for controlling the activation of knowledge sources. It developed plans for how to solve the alloy design problem by building context elements and high level goal trees. Once the meta space activated a context and goal, control was transferred to the appropriate domain knowledge source, which could build subgoals, develop the hypothesis tree or update the status of its goals to success or failure. The meta space could then build new context and goal elements based on the status of previous goals and the reasons for their success or failure. The meta level rules were formulated to reproduce the design strategy that was used in the aluminum lithium training case.

The hypothesis representation was also generalized so that decisions about processing and microstructure could be made as well as those about composition. This extension required that new links be invented to indicate the dependencies between decision nodes. Some of these dependencies were cause and effect relationships. For example, in order to meet a property target, the program would determine the required microstructure features and post those decisions as hypotheses. During a later iteration, the program would seek composition or processing options that were likely to produce the desired microstructure features. Another type of dependency came from multiple decisions that had to be selected together in order to achieve the desired effects. The first iteration hypothesis tree was essentially an "or" tree - all children of a single node were considered to be independent options. The second iteration representation allowed for "and" branches when multiple decisions had to be grouped together and treated as a unit. The restriction of building hypotheses using only depth first search was abandoned. A breadth first search of structure space, followed by an evaluation and pruning of poor options, followed by a depth first search of composition and process spaces seemed to be a better model of the human design process.

A general representation of multidimensional constraints was developed and used to implement a least-commitment decision-making strategy. Using this approach, decisions about exact amounts of additives or processing times and temperatures were postponed. Instead, the system derived linear or piecewise linear constraints in composition/ process space to define regions where properties were acceptable. As more property targets were considered, the sizes of these regions would decrease. If the initial problem was overconstrained or incorrect decisions were made during the search, conflicting constraints would be generated and detected. This would cause the system to backtrack or relax some of the property targets. On the other hand, if the system lacked sufficient knowledge of the problem (generation operators), or initial targets were too vague, the search procedure would terminate with a fairly large region remaining. In this case, the system would identify a family of points in the region and propose them for experimentation. In summary, the multidimensional constraints served four important purposes:

- Backtracking was reduced through the use of a least-commitment decision-making strategy.
- The interactions between search variables were represented.
- The constraint regions provided a criteria for success or failure of a design proposal.
- Multiple solutions (or one optimal solution) could be found for alloy design problems that lacked sufficient specification.

During the first iteration, design problems were decomposed according to property targets. However, most property targets were known to interact. In other words, operators that improved one physical property would also degrade other properties. Without taking into account these interactions, there was a good chance that the system would loop indefinitely. Since the nature of some types of interactions were known and understood by metallurgists, this knowledge was incorporated into the hypothesis-generation operators. These operators fired on the existence of an active goal to improve some physical property. However, the operators also checked for the existence of posted goals related to interacting properties in their preconditions. In this way, preference was given to operators that did not adversely affect other property targets being pursued in the same design problem. Other more subtle property interactions were detected during the hypothesis-selection context. Hypothesis-selection was preceded by an heuristic evaluation of all target properties. Selection was based on a measure of the distance between target and estimates in multidimensional property space. Hence, the effects of hypotheses on other property targets not being directly pursued in the current iteration were at least considered.

The second iteration design has been formulated and partially implemented. It is expected that implementation will be complete by the end of 1985. The alloy designers will then aid in another evaluation of the system.

During 1986, the ALADIN system will be converted to run on a Symbolics machine at ALCOA. After that, a more suitable user interface will be developed so that more alloy design experts will have easy access to the system. The Carnegie-Mellon team will complete its work and an ALCOA project team will prepare for continued development of the system. Major changes to the architecture or design are not expected, but a great deal of work remains in order to build the knowledge bank to an acceptable level. Also, the rule system must be generalized to deal with a broader class of problems. Several new training cases will be identified, and the reasoning will be characterized using the procedures established for the first training case. The rule system will then be augmented to reason correctly for all training cases. It is expected that the Artificial Intelligence group will play a gradually decreasing role on the project and that the Alloy Technology Division will increase the level of responsibility. This is because future work will require an ever-increasing level of knowledge of metallurgy.

5. Conclusions

Several insights were gained by our work on the ALADIN project. These insights pertain to knowledge acquisition techniques, design, architecture and implementation.

With respect to knowledge acquisition, it is very important to keep the teams small and give each individual specific responsibilities. Writting a detailed script of the knowledge and reasoning used on training cases serves several purposes and will be used on other projects. The script helps to uncover errors in communication before plans proceed too far, it helps the expert to understand some of the major computing concepts used on his problem and finally, it defines requirements during system design and implementation.

Some of the design ideas contained in ALADIN can also be extended to other systems. The use of control elements to decompose a production system into smaller units and to control rule firing has been suggested and used before [1]. The goal management subsystem is particularly powerful and is flexible enough to handle a variety of situations. Similarly, the idea of using multi-dimensional linear constraints to gradually converge on the solution to a problem with a reduction in the amount of backtracking can be applied to other application areas. The diagnostic checking rules have saved a great deal of debuging time and are recommended for complex systems requiring more than one developer. Finally, the rapid prototype method is a good one to use, and it is most effective when the prototype and the initial design are kept as simple as possible.

6. References

1. Brownston, L., Farrell, R., Kant, E., Martin, N., *Programming Expert Systems in OPS5 - An Introduction to Rule-Based Programming*, (Addison-Wesley Publishing, 1985).

2. Bryant, G.F., *Automation of Tandem Rolling Mills*, (The Iron and Steel Institute, 1973).

3. Buchanan, B.G., Shortliffe, E.H., *Rule-Based Expert Systems*, (Addison-Wesley Publishing Co, 1984).

4. Carnegie Group Inc, *Knowledge Craft Beta Release Notes*, (1985).

5. Foderaro, J.K., *The Franz LISP Manual*, University of California Technical Report, (1980).

6. Forgy, C.L., *OPS5 User's Manual*, Department of Computer Science Technical Report, Carnegie-Mellon University, CMU-CS-81-135, (1981).

7. Hayes-Roth, F., Waterman, D.A., Lenat, D.B., *Building Expert Systems* , (Addison-Wesley Publishing Co, 1983).

8. Hulthage, I., Rychener, M., D., Fox, M., S., Farinacci, M., L., *The Use of Quantitative Databases in Aladin, an Alloy Design System*, Robotics Institute Technical Report, Carnegie-Mellon University, CMU-RI-TR-85-19 (1985).

9. Kobayashi, S., *A Review of the Finite Element Method and Metal Forming Process Modelling*, Journal of Applied Metalworking, vol 2, # 3, (July 1982), 163-169.

10. Nilsson, N.J., *Problem Solving Methods In Artificial Intelligence*, (McGraw-Hill Book Company, 1971).

11. Steele, G.L., Jr., *Common LISP the Language*, (Digital Press, 1984).

Acknowledgements

We are grateful to the metallurgists who provided case histories of design problems for this project. In particular, we thank Marek Przystupa and Warren Hunt who aided in knowledge acquisition and design during the second year.

ARTIFICIAL INTELLIGENCE IN MANUFACTURING
T. Bernold (editor)
Elsevier Science Publishers B.V. (North-Holland)

C O N A D (CONfiguration ADvisor) LINKED TO ITS ENVIRONMENT

Bernhard Franz
Management Expert Systems Sales
Nixdorf Comupter AG, Munich/West Germany

C O N A D (CONfiguration ADvisor) Linked to its Environment.

About us:

In the information industry Nixdorf ranks among the four major European companies. Its activities center on services and system solutions designed to give users flexible and efficient support in their business operations. As one of its principal strengths, it utilizes its technology resources to meet market needs through user-orientated products.

As a company, we have a record of exceptional growth. Since Heinz Nixdorf's "Labor für Impulstechnik" founded in 1952 was renamed Nixdorf Computer AG in the year 1968, the number of employees has increased twelve times, revenue has risen 31 times and stockholders' equity 49 times.

Foil 1 "Revenue"

In the same period, the marketing and service organisation has consistenly expanded and now embraces a network of 41 countries. This extensive international presence allows us to focus on market requirements in individual countries so as to further future company growth.

Foil 2 "Corporate Structure"

In addition to this service organisation network throughout 41 countries, Nixdorf maintains seven production plants. These plants are in Germany, Ireland, Spain, USA and Singapore and form a manufacturing combine with approx. 6.000 employees.

Foil 3 "Production Plants"

The basis of the product-orientated communication between marketing, research, sales, field service, calculation and production is the so-called "Systemkomponentenschlüssel-Nummern-System" (SKS-System).
(literally: system component key number-system)

SKS numbers describe system components as devices, fragments or part of devices, connections etc. SKS-numbers can be found in sales price lists where they represent - phased out independently from systems - system models in combination as well as in work planning and control.

Foil 4 "Deutsche Preisliste"

Foil 5 "SKS Stuecklistenaufloesung"

Let me show you an example of our product family 8864, the Nixdorf system family for banks, the automatic teller machine as it shows up in the price list and bill of material.

Foil 6 "SKS Number"

At this point I think it will be helpful to explain the SKS processing theory in more detail, i. e. the conversion into bills of material.

The relevant bills of material are produced from the SKS by converting each position of the SKS number into one bill of material and then to combine it with the bills of material already found.

An example:

First, the complete SKS is broken down into its SKS positions.

The following scheme will explain the method by using the automatic teller machine already mentioned.

Each position within the SKS describes a specific component, the component number is used in order to address the respective modules in the data files and to add them together. This means the planning standards for production budget and production are considerably dependent on the correctness of the imported SKS numbers. The latter definitely have to be a correct and complete conversion of the customer's purchase demand.

In order to secure this part Nixdorf has created a tool for the SKS plausibility and quality check with respect to incoming purchase orders for sales and production:

CONAD, a configuration counselling expert system

The aim of the application is to use computers to enable sales personnel to generate custom-made purchase hardware configurations and production standards based on a price list structure.

Foil 7 "CONAD"

As a matter of course, the applicability of CONAD exceeds the aforementioned fields. Marketing and research that convert future marketing demands can be integrated in such a system as well as field service and price calculation.

Why does Nixdorf use their own expert systems?

Foil 8 "Applications at Nixdorf"

1. Company policy is to offer final consumer orientated solutions simplifying and rationalising customers' routine work, and to secure this Nixdorf runs home key projects.

2. Nixdorf wants to use the solution itself immediately and efficiently.

A real case:

Nixdorf takes pride in stating that they hold first place as partners of banks. And that does also mean that they hold first place in offering solutions - and not only software solutions but even complex hardware solutions. In addition to the central processing unit and peripherals, various aspects have to be taken into consideration when arranging a work station, i. e. screens, keyboards, printers, identifyers, automatic teller machines, automatic teller safes etc.

This kind of complexity often results in incomplete purchase dates which cause evitable inconvenience (even financial) to our customers as well as internally, i. e. unprecise order confirmations, delivery delays or partial delivery, product personnel cannot work normally because of customer inquiries or the dispositionary range is restricted by vague purchase data or non-scheduled re-ordering.

CONAD's knowledge base contains all the information necessary to construct each device, contains information on system components, their criteria and compatibility etc (taxonomy and rules).

Foil 9 - 18 "Konsultation gestartet...."

This basis enables the user to entirely configurate "his" computer system, in dialogue or controlled by rules. Furthermore, the explicability of an expert system ensures that the user can analyze the knowledge base (incomprehensible proposals etc.). In reply to commands such as

Why? Facts! How? Knowledge! Rule? Advice?

the expert system gives explanations on what knowledge base conclusion have been drawn and proposals made.

Foil 19 "Das/der Ver-/Entschluesselungsmodul"

My last scheme shows the final configuration of a lobby-type automatic teller machine based on data deduced from existing inquiries or purchase orders by the user.

Following the method initially described configuration data can be verified by means of CONAD. SKS numbers can be easily converted into components of bills of material.

Let me show you an excert of such a conversion.

Foil 20 - 24 "SKS Stuecklistenaufloesung"

Foil 25 "Applications at Nixdorf"

Once a consultation session has been completed "to the satisfaction" of CONAD it has been guaranteed that the Nixdorf system 8864 thus ordered is free of faults, that no part has been forgotten and that all parts fit together.

CONAD is applicable to all configurations based on modular components and has been built with the Nixdorf expert system shell TWAICE (registered trademark).

ARTIFICIAL INTELLIGENCE IN MANUFACTURING
T. Bernold (editor)
Elsevier Science Publishers B.V. (North-Holland)

SYSTEM DEVELOPMENT IN A LISP MACHINE ENVIRONMENT

Symbolics GmbH, G. Timmermann

The more complex the problem, the better the tools needed for efficient programming. LISP is the best choice for developing many types of symbolic programs, especially expert systems. It runs most efficiently on specialized hardware - LISP machines - , on which it runs in a sophisticated, highly integrated system development environment.

The area of expert systems may be regarded as a subdomain of artificial intelligence. Principles and solutions other than those used in traditional data processing are employed in this area. Mostly, a much higher level of abstraction is adopted through data structures and computing strategies which are closer to human thinking and problem solving than numerical processing. Of major importance are data of symbolic as opposed to numeric nature. Hence, the introduction of the term "symbolic processing" and the start of a new era in computing.

In the 70's we saw a hardware "paradigm shift" as mainframe vendors offered their computing cycles in more price competitive supermini class packages and supermini vendors succeeded in further reducing the size and cost of their conventional architectures by introducing personal versions of their hardware that ran the same operating systems and applications. Micro vendors started at the personal scale and successfully increased the power of their systems while keeping prices competitive. The net effect of the hardware "paradigm shift" is that the personal system is here as a host for the applications once run on systems that were cost prohibitive for the individual user. This trend will continue.

The Lisp Machine is an important entry into the field of computing. Lisp machines support the "paradigm shift" in software that is underway in the 80's. This new approach can be explained as follows. The programming techniques that have resulted from research in building artificial intelligence systems have been codified and are being used more and more by systems programmers to build advanced applications. These techniques include new ways of representing data in the form of knowledge that can be reasoned about. Also are techniques for manipulating the data symbolically, algorithmically, as well as logically. Programming styles known as object oriented programming, rule based programming, metaprogramming (embedding new languages and control structures in the host language), data driven programming and logic based programming are being widely used. These techniques make it possible to solve more complex problems and permit systems programmers to build more advanced man machine interfaces. These techniques have been used most frequently in what have been called Artificial Intelligence (AI) programs. AI applications have traditionally required large virtual memory to support their execution. In addition, the programs are data intensive, making use of various forms of data representation that are manipulated dynamically to produce desired program behavior. AI applications have demanded more from the environment of the computer being used to support their development. An environment for interactive symbolic software development and software-reengineering has evolved to support AI applications primarily as an example of an extensible symbolic system.

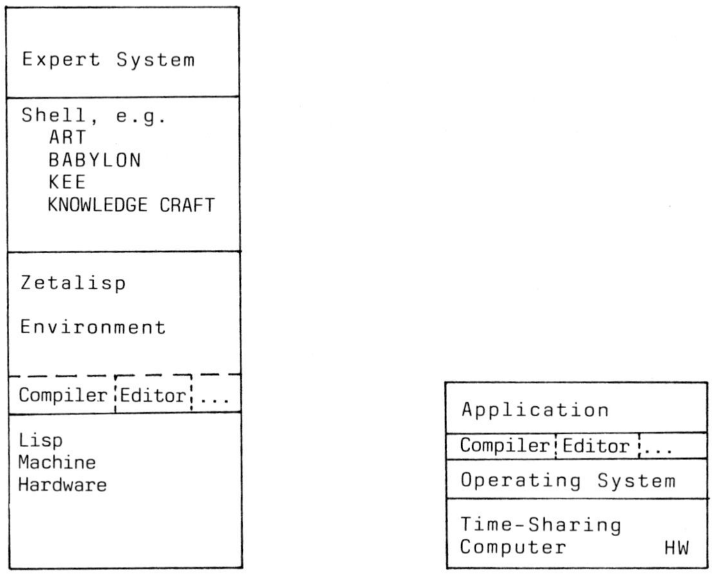

Expert System Design on LISP Machines

Traditional Approach

Diagram 1

Diagram 1 depicts an overview of the levels normally encountered in building expert systems.

As can be seen in diagram 1, expert systems are normally based on a special tool for building them called SHELLS. Some shells are commercially available, e.g. ART (Inference Corp.), BABYLON (GMD), KEE (Intellicorp), KNOWLEDGE CRAFT (Carnegie Group). All shells mentioned contain a number of different rule processing mechanisms which may be selected and merged as appropriate. During the last few years the complexity of human thinking has been recognized as using numerous strategies, therefore, isolated models like Prolog (efficient where usable, but backtracking with depth-first search only) or OPS 5 (forward chaining only) are very restricted in their range of applications. These newer tools have therefore integrated a number of different strategies as well. As an example of the requirements, look at the approach taken by the ALADIN system (paper given by Mrs. Farinacci from Alcoa Labs.) where CRL, OPS5 and LISP functions are used in conjunction.

What is it that expert systems require and traditional computer methodologies don't offer? Every experienced programmer knows that during the development of complex programs much time has to be spent debugging and testing. From a methodological point of view, the use of stepwise refinement has become very important. The development of expert system rule bases demands stepwise refinement. Therefore, the edit/test/debug cycle should be as fast as possible. When doing simulations and factory control the usage of multiple windows on a fast, high resolution screen as well as mouse-controlled menus are important ingredients. Data structures needed for expert systems normally include lists, trees and graphs. There are a couple of special functions provided to operate on those data structures, for example pattern matching.

Recognizing these problems, the AI group at MIT built an integrated development environment based on the programming language LISP. As they found that traditional hardware could accounter for the software demand only very little and with unacceptable runtime, they proceeded to build a specialized machine. The result of this approach was something like a "symbol cruncher" - in contrast to "number crunchers" in traditional Electronic Data Processing.

Zetalisp Software

Zetalisp is a dialect and improvement of LISP, furthermore it is a complete, highly integrated program development environment. LISP was characterized by Prof. GOOS /GOOS 1984/ (translation by the author):
"Most programming systems for tasks in AI are written in some dialect of LISP. This purely applicative language based on the -calculus has been blown up to its actual form by adding a large number of elements from imperative languages. Besides its often really good programming environments the unique suitability of tree-like LISP-lists for the representation of data structures used in this area contributes to its popularity as well as the possibility to interpret lists computed by the program as part of exactly this program; that is, LISP programs are not static but may change and expand themselves during execution."

For programming numeric or systems programs LISP is about as easy to use as a conventional language. For programming an extensible symbolic system such as an expert system, LISP has a productivity advantage.In building an advanced AI system, with a conventional language one encounters the "wall effect." The requirements of such an application make it nearly impossible to program it with a conventional language, no matter how much effort is invested. It is for this reason that Lisp machines were originally developed. It is for the same reason that the Lisp machine is becoming a popular industrial computer resource.
Now let's turn to the architecture of the machine to see how it works and to further understand how it is applied to complex problems.

To begin, the LISP language has evolved over a 25 year period to incorporate the best features of standard programming languages as well as to offer a set of unique features of its very own. LISP is unique in its uniform simple syntax and its common internal representation for programs and data. Many features such as list management, macros, and type and boundary checking are implemented more powerfully than in conventional languages on conventional machines. Other features such as object oriented programming extensions have a unique and powerful embodiment in the Lisp Machine architecture. For example, list management with symbols and properties are implicit data structures in the language and are directly supported in the hardware as are the structures required for object oriented programming. Macros obey the simple LISP syntax and hence can be compiled for more efficient execution. Complex data structures such as infinite precision numbers, arrays of mixed data types, arrays with infinite bounds (planes), and full user defined data typing make it possible to build sophisticated systems programs. As previously mentioned, LISP has conventional features such as structured programming, numerical operations, I/O processing, bit manipulation, etc. It is important to remember that the Lisp Machine not only proves efficient interpretation and compilation of LISP but also supports an environment comprised of over ten thousand functions that aid the system programmer in programming efforts including powerful text handling functions, searching and sorting routines, protocol management routines, etc. The SYMBOLICS 36xx family of computers offers a number of compilers for conventional languages such as Fortran, Pascal and Prolog. Not only is LISP good for embedding other languages but is also known as the machine language of high level languages. These compilers take full advantage of the Lisp Machine environment such as debugging tools, window system, etc. and are offered with their own language directed tools such as symbolic debuggers and editors. Offering conventional languages the same address space as LISP permits the melding of AI systems programming features with conventional application features. All of these software capabilities are provided in what is known as "a single level" memory environment. This enables the development and execution of programs with as much as 1 GB of virtual address space provided on the 36xx family of computers without concern for physical memory management.

The most important tool of every programmer is the editor. The ZMACS Text Editor from SYMBOLICS is based on Zetalisp and may be used in a unified way as well for LISP-, FORTRAN-, PASCAL- or PROLOG-programs as for traditional text processing. ZMACS makes full use of the window techniques and mouse- and menu-facilities that ZETALISP offers; it is possible to work on different files at the same time, as well as invoke one of the compilers from within ZMACS allowing quick debugging. In LISP mode the editor knows about the LISP syntax and supports the editing process by automatic syntax-driven indentation, highlighting of matching parentheses and many other syntax-oriented cursor commands. When set to some other programming language, ZMACS supports the editting of that language in a similar way.

The most significant feature of Lisp Machine is dynamic linking, that is, single functions or complete files may be compiled - or with LISP and PROLOG optionally evaluated - and immediately thereafter executed without any link/load run and without leaving the editor. A typical edit/compile/test-cycle is thus reduced to the following steps:

- editing using ZMACS,
- compiling from within ZMACS,
- test run from within ZMACS,
- if necessary, correction of a few modules and recompilation

of these, then the next test run.

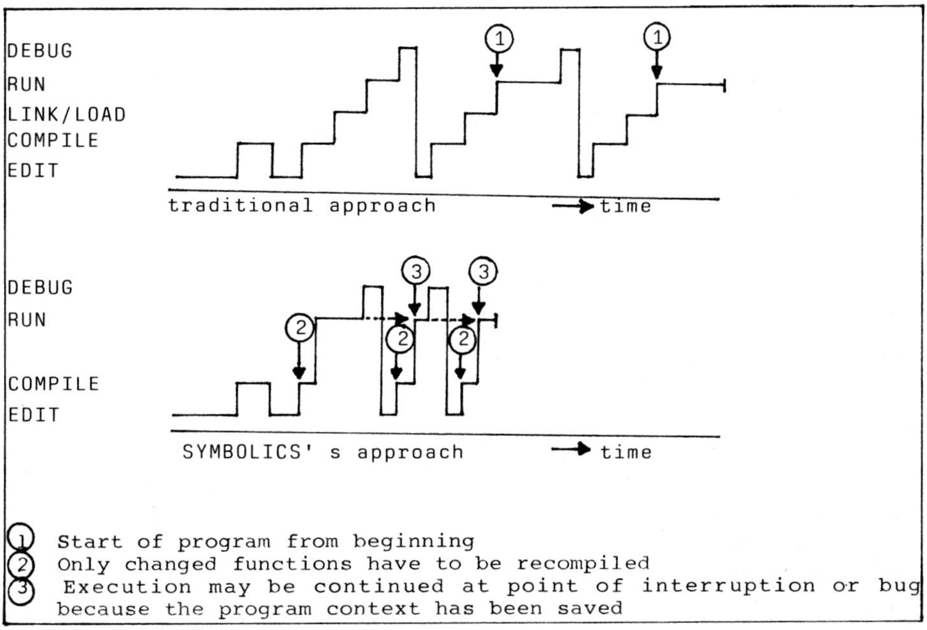

Diagram 2: Saving time during program development by omission of link and load, incremental compile and integration of tools.

In contrast to the traditional approach there is no link and load phase. This is an important prerequisite because it allows to test the result of small changes to large programs without the usual long delays.

Zetalisp has a special means for object oriented programming called flavors. The flavor concept is used in the system code heavily, and is an important tool for the user as well. Frames and similar concepts in expert system-shells are normally based on flavors. It's very important to note that the operating system itself is written in ZETALISP. The user has access to the sources so as to allow efficient use of all system facilities. Furthermore, adjustments to them can be made by direct modification of the system code.

For error tracing Zetalisp provides an intelligent debugger which automatically switches context to the language used. The debugger makes full use of the window system as do other tools as well. Whenever a runtime error is encountered, the debugger takes over. There is neither special preparation needed to use the debugger nor any runtime overhead as all checking is done in hardware in parallel to normal execution. The debugger not only reveals information about the kind of error and where it occurred, but also displays the number type and actual value of parameters of all functions in the calling hierarchy. In addition, the debugger may offer few options how to proceed, the number and types of those options being context dependent. To give a simple example: When dividing by zero the debugger provides the user with the option of either changing the value of the divisor or substituting the result of the devision directly. Another example: If an uninitalized variable is referenced the debuggger will offer to set its value. Of most importance is the possibility to call the editor from the debugger, correct a program part and then continue with the interrrupted program without losing intermediate results or parameters. It's even possible to freeze an error contextwithout any problems while developing and testing other programs thereby possibly using the debugger again - the different contexts will remain well separated. The debugger also offers such facilities as stepwise execution, trace, inspection and modification of the stack. Of course, the newly created standard COMMON LISP is part of the complete software package and may be used either restricted to the standard or improved by the flavor system for object oriented programming.

LISP Machine Hardware

The above-mentioned sophisticated and highly integrated tools are possible only by assistance of the underlying special Lisp Machine hardware. First, there is the size of the virtual memory, a minimum of 60 MB for a single user, up to more than 1 GB, being managed with demand paging. Splitting this memory into packages and areas results in a high level of locality of programs and data. The fast execution of code is possible through special hardware features for LISP, a large microcode (8 or 16 Kwords with 112 bits each) and special registers and caches for different types of LISP structures. The free storage management (patent applied for) runs in parallel and is assisted by hardware.

Another feature of the Lisp Machine is "data tagging," which means that data type checking is done by hardware in parallel to normal execution. This allows for generic operations; for instance, there is only one multiply operation for both integer and floating point numbers. It's the microcode that detects at runtime which types of data are to be multiplied - and branches accordingly. A special data tag driven, compact representation of lists not only reduces memory needs to a large extent, but in addition yields much higher execution speeds.

Integration into the EDP environment

Symbolics Lisp Machines are single user systems, which may easily be integrated into existing networks, either via serial lines or -more convenient- via Ethernet using IP/TCP, DecNet or CHAOSNET protocol. These high level protocols allow for transparent file access, mail exchange, remote login and other services to all connected machines, may they befrom SYMBOLICS or from other manufacturers. As SYMBOLICS provides an open system even user specific network protocols may be easily implemented.

Conclusions:

Lisp machines were built from the beginning to efficiently support a

sophisticated program development environment based on LISP, a highly symbolic programming language. For this reason, they are used heavily in the areas of AI and expert systems.
It does not make sense to use time sharing machines or traditional PC's for complex expert system development as these cannot offer the performance and level of integration needed for real-life problems. For systems development, one should not be satisfied with less than what current Lisp machines offer. With the finished products it may be different though: probably these may run on a somewhat less expensive workstation. Lisp machine vendors are currently working on those lower-cost, high-performance delivery vehicles.

<u>Literature:</u>

Goos, G., Programmiermethoden der Künstlichen Intelligenz in der Software Technik, in: H. Morgenbrod/W. Sammer (Hrsg.): Programmierumgebungen und Compiler. B.G. Teubner, Stuttgart.

DESIGN AND SIMULATION

ARTIFICIAL INTELLIGENCE IN MANUFACTURING
T. Bernold (editor)
Elsevier Science Publishers B.V. (North-Holland)

DESIGN FOR ASSEMBLY: A CHALLENGE FOR EXPERT SYSTEMS

Dr.-Ing. Arun Gairola

Digital Equipment International GmbH
Kaufbeuren, FRG

Anticipating the increasing use of new and emerging computer technology in product engineering, design for assembly will be achieved effectively by applying advanced data processing technology and data management methods creating an assembly knowledge base and decision support system to close the loop between design, process planning and actual assembly. This paper introduces DFA and describes a strategic and conceptual approach to realise such a decision support system leading to an expert system for Design for Assembly in the near future.

1 INTRODUCTION

The current approach of assembly optimisation - particularly through automation - often fails because of inadequate design of products with respect to their assembleability. Past experience, in assembly automation projects in particular, shows that design changes were necessary to allow technologically and economically successful automation. As shown in Fig.1 an investigation of several existing and newly designed products revealed that nearly 60% of parts required design changes to facilitate automated assembly[1]. The most interesting and significant result of this investigation is the fact that the majority of cases required only minor design changes not affecting the functionality of the product.

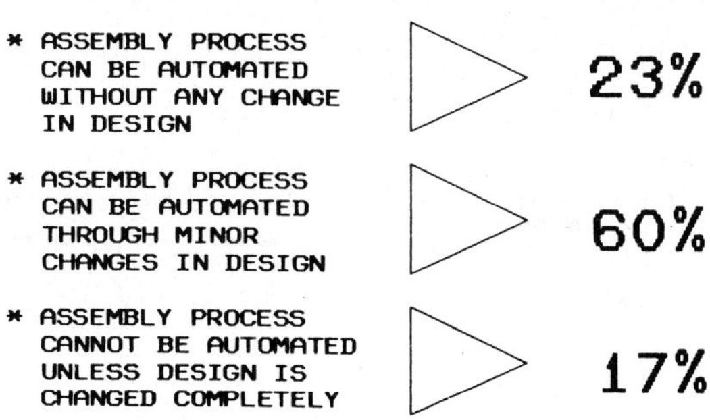

Fig.1
Result of Evaluating 10 Products regarding their Assembleability

Design For Assembly (DFA) is the technique to design products with ease of assembly in mind. By applying DFA a product can be systematically designed to minimise the technological and financial efforts required for assembly while still remaining within the constraints of the product's functionality. Fig.2 illustrates some examples of DFA.

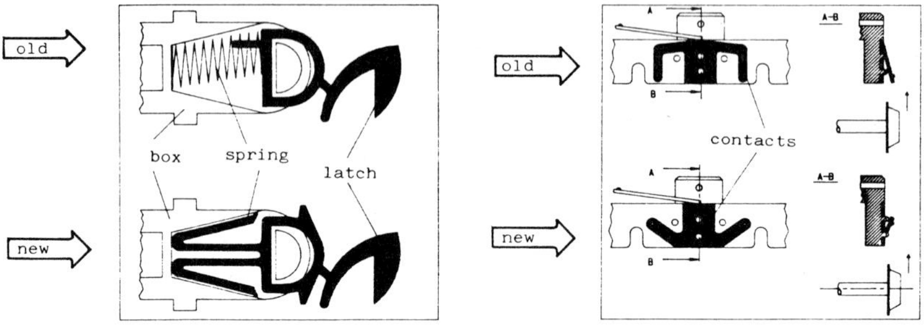

Fig. 2: Examples of DFA

* Cost saving
* reducing rework
* increasing quality
* minimising material cost

inside the industrial production system requires access to to methodological approaches, which enable the product engineer to design products with optimum ease of assembly (high assembleability). In view of the fact that today designers have to process a huge amount of information just to guarantee the functionality and the operation of the product as shown in Fig.3, it seems unfair to put further demands on the already overburdened designer. Therefore, to achieve the above mentioned goals a DECISION SUPPORT SYSTEM based on the knowledge of assembly requirements is needed to help the product engineer to optimise

* design for easing manual and automated assembly
* design for assembly process planning
* design resulting in less reworking and high quality

The main objective of this approach is to provide tools and methodologies that require minimum efforts from the designer to create such designs.

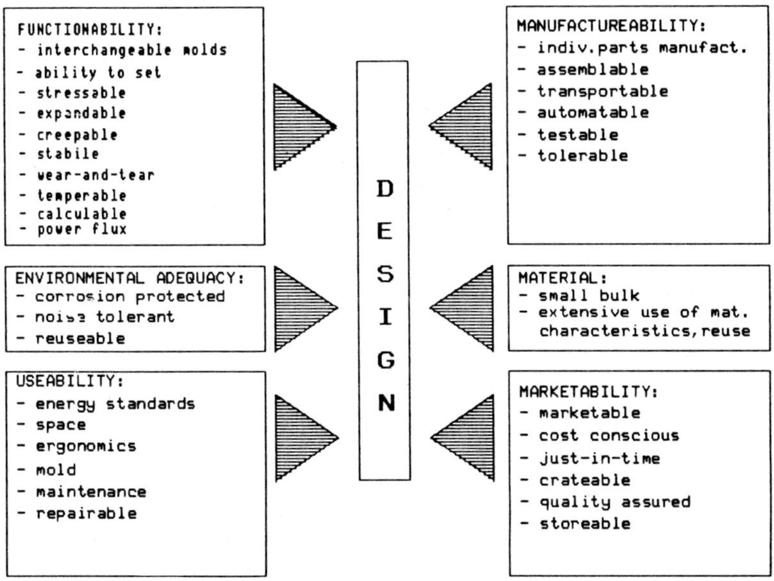

Fig. 3 Constraints on Design

2 PRACTICAL SIGNIFICANCE OF DFA

The ideal approach of DFA would be to completely avoid the need for assembly. In this case the product would consist of a single part. However, in the real world this approach is unrealistic because there are a number of reasons why a product must incorporate many different parts. The four major reasons are:

< to ensure the functionality a product must have parts made of different materials or parts which move relative to each other, i.e. to design a specific rotating motion a shaft must move relative to bearings

< to achieve low product costs without affecting the functionality of a product, parts must be made of different materials, i.e. use of synthetics in conjunction with metals

< to allow accessibility of assembly tools or other parts to already assembled parts, e.g. a cabinet generally consists of at least two parts to facilitate functional parts enter the cabinet

< to permit servicing of the functional parts by disassembling a product.

In spite of these constraints there are a number of ways to reduce the number and variety of parts in a product and to ease the execution of the assembly process. Design for assembly is concerned with design features that help simplify the overall assembly process, particularly for automation. In fact nowadays design for assembly is commonly regarded as design for automation. It is true that a product that is designed to ease automatic assembly is also easier to assemble manually. Although automation plays an important role, DFA is more than just design for automation, as illustrated in Fig. 4. All together there are three fundamental approaches to designing a product for assembly:

* Simplifying of assembly task

* Improving assembly organisation

* Facilitating assembly execution

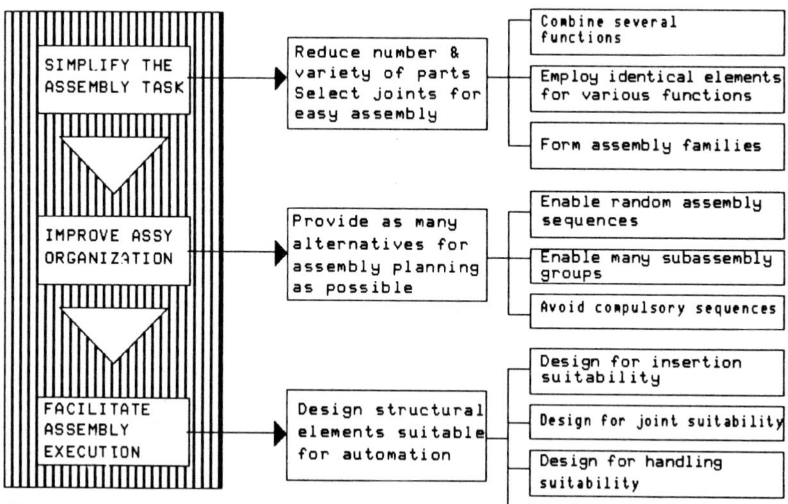

Fig.4 Requirements for designing for automated flexible assembly

2.1 SIMPLIFICATION OF THE ASSEMBLY TASK

The main goal of simplifying the assembly task is to reduce the content and the complexity of the task by

* minimising the number of parts
* reducing the variety of parts
* selecting simple assembly methods

The objective of simplifying the overall assembly process by reducing the complexity of certain assembly tasks is attained by minimising the total number of different parts in a product. Ease of assembly is a major objective of DFA and its impact is far reaching. A reduced number of parts will mean

* low inventory
* reduced material flow
* less number of work-stations
* decrease in number and variety of tools
* less production space
* cut down in planning efforts

There are basically two fundamental approaches to reduce the number of parts in a product. One approach is to determine the theoretical minimum number of parts needed to guarantee the product's functionality. Experience with several practical examples proves that the theoretical minimum number of parts can be determined effectively simply by examining each designed part with respect to whether it is required to ensure the functionality of a product. If a part has to meet any of the aforementioned fundamental criteria for designing a product composed of many parts then it must be a separate part.

The second approach to reducing the number of parts is their functional integration. Even if different parts have to fulfill one of the assembly criteria, in many cases it is possible to integrate them into one part by using specific materials and technologies. The prime example of this kind of integration is the use of synthetics in combination with metal parts in the OUTSERT and INSERT technique. In the outsert technique different functional parts made from synthetic material are integrated in a metal chassis through injection moulding technology. In insert technique, on the other hand, different functional parts made from metal are embedded in a base component by ultrasonic insertion.

However, the integration of parts in both cases theoretically means an increase of costs in the part's fabrication. In such instances a cost justification model is needed to evaluate the overall advantage of integration. But in the practice it has been proved that in most cases such an integration reduces the total cost of parts in spite of the increased complexity of the integrated part.

A large variety of parts requires a high degree of flexibility in the assembly system which could be outside the limitation of economic and technological feasibility. A possible way to avoid this is to design parts with common assembly features. It is not necessary to design parts with communality in every respect of assembly. Sometimes a very small number of similar characteristics makes it possible to build assembly families that can be processed with single assembly technology. Fig. 5. shows examples of such families based on specific common characteristics of the parts.

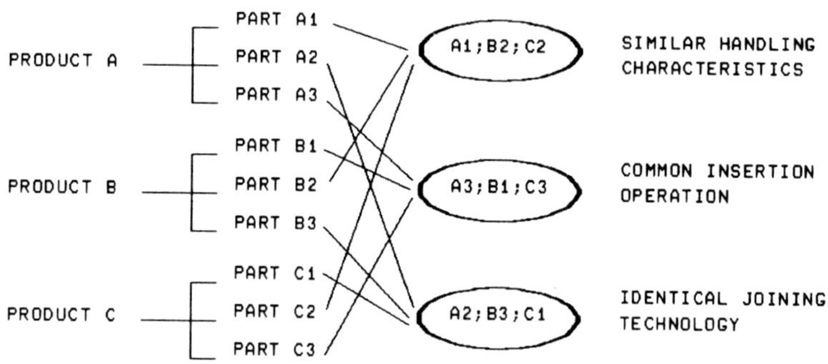

FIG.5: BUILDING OF ASSEMBLY FAMILIES

2.2 IMPROVEMENT OF THE ASSEMBLY ORGANISATION

The three major approaches to improving planning and performance in assembly through specific design characteristics are as follows:

* allowing arbitrary decomposition of the product in subassemblies

* using standard parts and subassemblies

* avoiding compulsory assembly sequences

Dividing a product into subassemblies is required because many products are too complex to be assembled in a single assembly process. Moreover, it is often necessary to use different assembly technologies, which are installed in different locations, and it is sometimes necessary from an economic point of view to standardise subassemblies. Standard subassemblies can be used in different types of product having similar kinds of sub-tasks,thereby reducing inventory levels and increasing production volumes. Thus, for assembly planning it is more efficient to decompose a product into small subassemblies, which can be independently processed.

To solve the problem of line balancing - a major task of the assembly planning - it is extremely convenient for the assembly process to be free from sequencing constraints. Free sequencing of the assembly process means that if parts A, B, C,D, and E belong to a product then the assembly process can begin with any of these parts and continue in any direction (see Fig. 6).

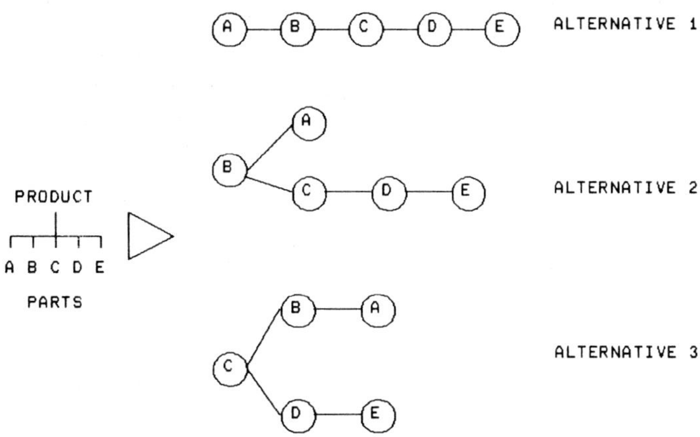

Fig. 6 Alternative assembly sequences

2.3 FACILITATING THE ASSEMBLY EXECUTION

Assembly comprises a large number of operations that can be classified into the following four groups:

* insertion

* joining

* handling

* controlling, inspection, and adjusting

For insertion of parts, particularly in case of machine insertion, it is important to design products in such a way that parts can be inserted successively downwards from the top. Product designs based on the so-called sandwich principle are typical examples of this kind of assembly process. Investigations of 355 insertion processes have shown that forcing a machine to insert from the side would be twice and from the bottom three times as expensive as from the top downwards. Results of this investigation are illustrated in Fig. 7. For this evaluation almost twenty criteria were considered, e.g. planning costs and time, technological efforts, machine down-time and reliability, utilisation, productivity, costs of maintenance and repair, etc.

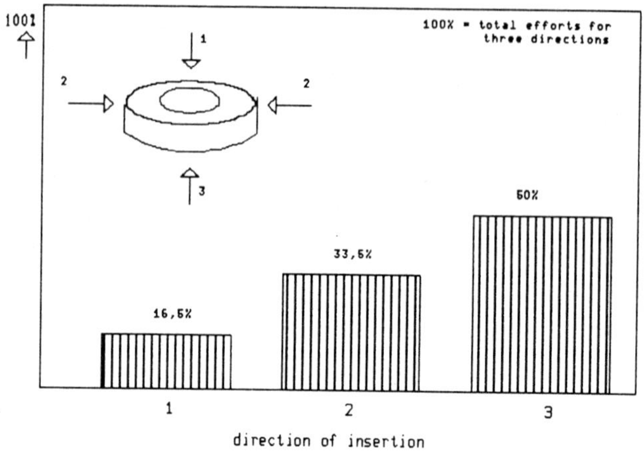

Fig.7 Insertion Directions and realization efforts

The main task of joining is to prevent the mated or inserted parts from falling apart. Execution of this process can be simplified by using joint technologies which do not require any additional parts, like nuts and bolts, or other work intensive processes as is the case with gluing. Further simplification can be achieved by using linear joint which require insertion along only one axis and either use no specific tools or require only standard tools. The best solution in this case would be to use joint components which can themselves act as a tool.

Parts handling consumes almost 80% of the assembly time. The maximum savings in terms of costs and time can therefore be obtained by avoiding or simplifying the handling of parts and subassemblies. The ideal way of doing this is to design parts with additional specific form elements which actively support the execution of the handling process.

Control operations during and after assembly are needed to check the quality consistency of the assembled parts relative to their functionality. A reduction in the complexity of this operation can be attained by designing components to facilitate inspection and allow free access of tools to test and adjust points.

In a comprehensive research project almost 160 design rules for DFA based on above-mentioned thoughts have been formulated and compiled in a "book-of- rules". In an appendix this book-of-rules contains approximately 400 illustrated practical examples of successful application of the design rules on parts and subassemblies.

3 METHODOLOGIES FOR DFA

3.1 STATE OF THE ART

To accomplish an adequate design for automated assembly the product designer must have a profound in-depth knowledge of assembly techniques and be familiar with current and future assembly technology. In view of the large range of requirements with which a design must comply and the magnitude and diversity of the fields of knowledge required to fulfil these demands it seems impossible for a designer to meet this challenge using conventional methods of design. This situation becomes even more critical due to constantly decreasing product life cycle. Within the tight design schedules designers cannot afford to check all possible design characteristics not related to the product functionality.Thus, in practice, the designer checks a design only with respect to its functionality so that assembleability is only a minor consideration. It is virtually impossible to redesign the product to improve assembleability after the design has been finalised. In view of the importance of well-designed products with high assembleability to reduce the total product costs, a tool must be developed to preevaluate a product during the design phase whether it can be assembled with minimum efforts. The tool must be designed in such a way that it also helps the designer to establish appropriate design modifications if any flaws are detected.

A tool of this kind could take the following form:

1 Parallel to the design process the tool analyses all decisions relevant to assembly made by the designer and points him to the faulty decision.

2 If necessary, it indicates the critical features in the design.

3 If necessary, it proposes possible changes of the inappropriate design characteristics.

4 To help find a suitable solution it provides examples of design changes. The tool will not be able to provide a unique solution to the specific problem because a design change must be checked to see if it impairs the product functionality. The designer of the product is the only person authorised to do this.

Many experts working in the field of DFA have recognised the need for such a tool, and they have developed a number of different tools to help designers conceive products that are consistent with the objectives of DFA [2], [3], [4]. The main problem in using all these tools is either that they lack entirety in dealing with the assembly problem, or that their application is tedious, cost intensive and extremely time-consuming. Moreover the application of these tools does not guarantee success in every case because most of the knowledge on DFA exists only as experience which brings irreproducible results. At present there very few approaches which attempt to formalise available knowledge in the field of DFA to achieve reproducible results [5], [6]. The results are still not satisfactory.

3.2 DFA AND EXPERT SYSTEMS (ES)

In view of the increasing use of new and emerging computer technology in product engineering the application of DFA can be facilitated to a great extent by applying advanced data processing technology and data management methods. Extracting knowledge on DFA from a human expert and putting it in computable forms can greatly reduce the time required for knowledge reproduction and exploitation. Examples of such applications are today well known expert systems. Unlike traditional data processing systems, expert systems involve many distinguishing features, which include the symbolic representation of knowledge, and symbolic inference based on heuristics.Heuristics is a domain of science that helps to solve those problems not accessible by formal techniques.

The basic characteristics of ES are:

< performing difficult tasks at expert levels

< emphasising domain-specific problem-solving strategies

< applying selfknowledge to reason the inference process

< providing explanation or justification of the conclusions reached

< involvement of paradigms, tools and system development strategies

Expert systems are characterised by specific names depending on their domain and mode of application. Decision Support Systems are Expert Systems utilised to support a user in the decision-making process (see Fig. 8). The underlying philosophy of DSS is that computer technology is extremely useful in improving and expediting decisions, particularly when a large amount of information or data has to be processed. In the case of DFA this can be done by preprocessing data and expediting informational flow, explicit models to provide structure to the decision process and explicating known examples. Fig. 9 shows seven different modules necessary to encompass the complete domain of DFA objectives.

Knowledge is the key ingredient in DSS. The efficiency and scope of DSS is more dependent on the knowledge it possesses than on the particular knowledge representation method and inference scheme it employs. Expert's knowledge provides the key to expert performance, while knowledge representation and inference schemes provide the mechanism for its application. Therefore, the strength of an expert system derives the knowledge it possesses and its quality from the particular formalism or inference schemes it employs. The knowledge of an expert is essential for the development of a decision support system and almost sufficient if available formulated in a logical form.

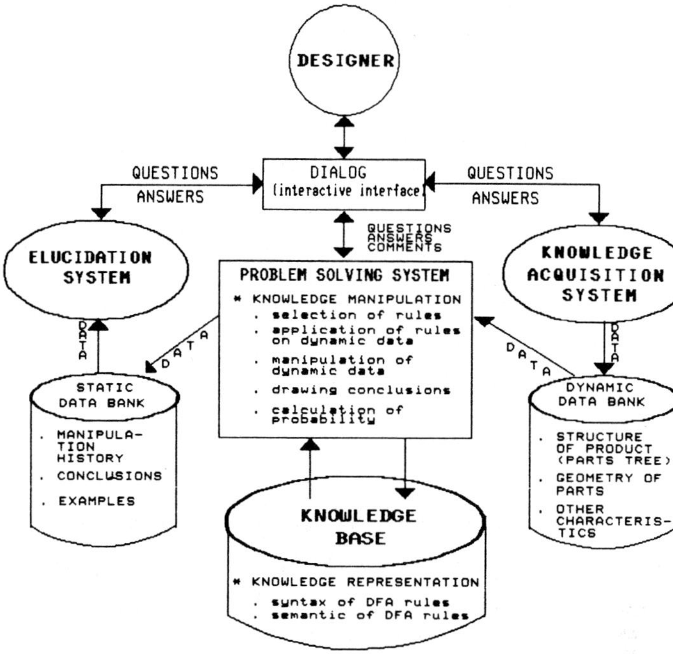

Fig.8 Information and Data Flow in an Expert System for DFA

Extracting, articulating and computerising the expert's knowledge constitute the essential tasks of developing DSS. The knowledge in a given domain of interest consists of descriptions of formal facts and empirical associations, relationships between these and procedures for their application. These relations express dependencies between items in the knowledge base of a DSS. In the case of DFA most of the knowledge required is now available in a well formulated form, which thus solves the greatest problem of extracting and collecting the knowledge from different sources. This prerequisite alone makes DFA an ideal field for the application of expert systems as a decision support system. Going through the checklist for selecting problem domains where DSS can be used, DFA meets anyhow virtually all the requirements shown in Fig. 10.

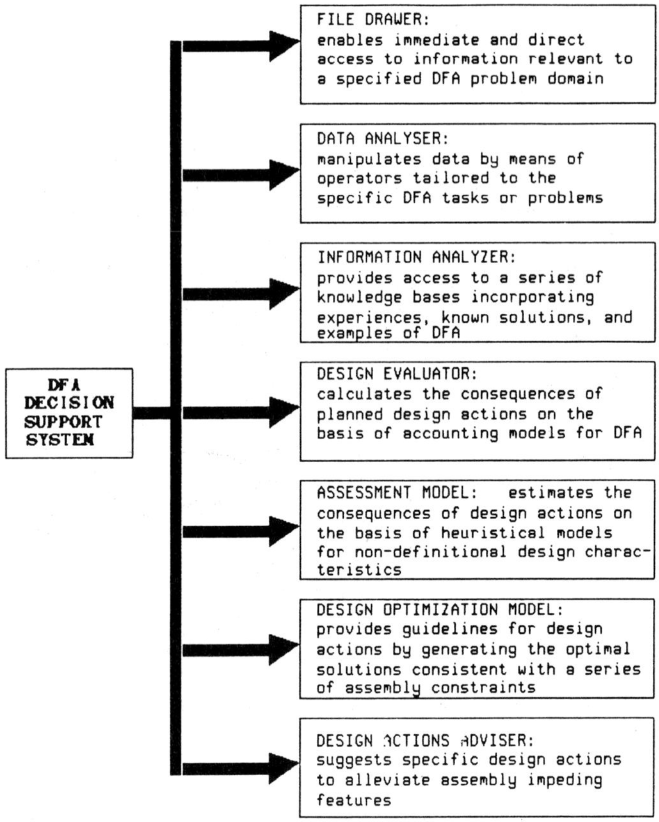

Fig.9 Elements of DSS for DFA

4 CONCLUSION

Experience with DFA to date shows that the designer is not able to take the assembly requirements into account adequately. The reason for this is the great variety and diversity of information which has to be processed in a short period of time by the already overburdened designer. The complex interrelationships between the design features and their influence on the assembly complicate further the DFA process. It is necessary therefore, to provide the designer with a tool which can help him design products that are appropriate for assembly. To meet this challenge, recent developments in computer technology, particularly in the field of expert systems, can make a significant contribution. DFA provides the ideal preconditions for the application of expert systems.

1.	Can the problem domain be defined very clearly?	yes
2.	Does the problem require expert knowledge and judgement?	yes
3.	Are there competent experts or recognized expertises available?	yes
4.	Is there a significant demand for the solution?	yes
5.	Is conventional programming not applicable due to intractable problem?	yes
6.	Does the problem require the use of heuristics?	yes
7.	Is the amount of knowledge required to solve the problem very large?	yes
8.	Is the problem domain fairly stable?	yes
9.	Is it acceptable to have solutions with certain probability (not all or no solutions)?	yes
10.	Is the expertise based on experiences over a long period of problem solving?	yes
11.	Would potential users welcome the system?	yes
12.	Are the benefits of the system accountable?	yes

Fig.10 Checklist for selecting problem domains for DSS

5 REFERENCES

[1] Gairola,A. and Voigt,B., Massnahmenkatalog fuer das montagegerechte Konstruieren, VDI- Berichte 556, VDI Verlag Duesseldorf,1985

[2] Gairola,A., Montagegerechtes Konstruieren: Grundlegende Gestaltungsmassnahmen Technische Informations - Bibliothek Hannover,1984

[3] Gairola,A.,Design for Automatic Assembly,in Warnecke,H.J.and Bullinger,H.J. (eds), Factory of the Future (Springer,Berlin,1985)

[4] Andreasen,M.M.,Kahler,S. and Lund,T., Design for assembly (IFS publications Ltd., Bedford, UK)

[5] Gairola,A, Montagegerechtes Konstruieren: Ein Beitrag zur Konstruktionsmethodik, Ph.D. Thesis, University of Darmstadt, Darmstadt, W. Germany 1981

[6] Boothroyd,G. and Dewhurst,P., Design for assembly - a designer's handbook, University of Massachusetts, Amherst, USA

ARTIFICIAL INTELLIGENCE IN MANUFACTURING
T. Bernold (editor)
Elsevier Science Publishers B.V. (North-Holland)

FROM CONCEPT TO PRODUCT: CREATING AN EXPERT TOOL FOR THE DESIGN OF AUTOMATED FACTORIES

Robert C. McArthur
Artificial Intelligence
Arthur D. Little, Inc.
Cambridge, Massachusetts
U.S.A.

Recently Arthur D. Little, Inc., undertook an assignment to evaluate the applicability of AI technology in the design and installation of automated materials-handling facilities. As a part of this assignment, a prototype knowledge-based system was built to support the construction of facility design during negotiations with a potential customer. The primary objective of the prototype was to allow rapid specification and modification of a given design, as well as to provide a facility for eliciting the expert judgement and hueristics used by the client's systems engineers. The project approach and prototype operation are discussed.

BACKGROUND

Recently, Arthur D. Little, Inc., (ADL) was asked to assist a client in constructing a prototype knowledge-based system to support their ongoing operations in the area of automated materials handling systems. These systems are known in the trade as ASRS. While the client was very sophisticated in his use of computer technology, his organization was not familiar with the use of artificial intelligence techniques. The client was, however, convinced that AI technology would eventually change the basis of competition within his industry, especially in the area of ASRS design, and was eager to begin developing internal expertise in the use of that technology.

As we began our work, it was not clear what the first application of AI technology should be in the client's organization. Thus, in addition to designing and implementing the prototype, part of ADL's task was to identify the most appropriate application area. After discussion with our client, it was decided to limit the search for application areas to problems relating to design and installation of ASRS facilities.

The ASRS industry has seen a radical shift in technology during its short lifetime. In the early 1970's, ASRS suppliers were largely concerned with the speed and reliability of their hardware, and the relay-based control logic that was used was not difficult to design. By the mid 1980's, however, the situation was reversed: ASRS hardware was both reliable and widely available, while the design and testing of ASRS software for control was becoming a major part of the total cost to develop ASRS facilities.

Rapid increase in business volume during the previous several years had put our client into a somewhat precarious position, for while he employed a sufficient number of systems engineers, the vast majority of these engineers were newly hired and had little practical experience. Thus, relatively few senior-level systems engineers were available, and they were spread thinly over many projects. The lack of technical leadership caused critical bottlenecks to occur, both in the execution of existing contracts and in the development of new business. It was hoped that AI technology could help ease the burdens of senior staff by allowing junior staff to perform some of the tasks that required senior engineers.

During this period, the rapid change in ASRS design requirements created another problem: a difficulty in establishing common design criteria and functional specifications for hardware and software designers. In many projects, in fact, considerable effort was devoted to the transfer of knowledge from one group to the other during the design process. This knowledge transfer, then, had become an additional bottleneck which our client was actively trying to overcome, and there was a general consensus that AI technology had something to offer in this area as well.

APPLICATIONS CONSIDERED

The first part of our project involved an exploratory effort to uncover the most cost-effective applications of AI technology within the client's organization. During the effort, three principal application areas were seriously discussed:

- automation of the design process including final optimization;
- diagnosis of problems in newly-installed ASRS facilities; and
- technical assistance during negotiations with potential customers.

The principal thrust of all of these applications was to reduce the workload of the organization's senior systems engineers by allowing some of their responsibilities to be carried out by more junior-level individuals.

Design optimization was considered a task which, although tedious, was too difficult to be left to junior engineers. Because of the complexity and non-linear behavior of most ASRS designs, it was nearly impossible to optimize a given design mathematically. This left as the only alternative a long process of trial-and-error during which the design would slowly approach satisfactory performance and cost. Senior engineers were able to work much faster and converge to a higher level of performance.

The troubleshooting of newly-installed facilities was similarly tedious, but critical to the overall success of the organization, because it was during this period that the newly-installed facility was tested as to whether it met contractually-specified performance requirements. If the installation and testing was especially difficult and/or time-consuming, there was a good chance that the project would overrun in cost, or perhaps never reach its target specifications during the allowed time period; either of these possibilities was unacceptable.

The providing of technical support during customer negotiations was seen as the area in which the organization needed the most immediate assistance. First of all, the preliminary design given to the customer had to be based on customer-supplied information which was often spotty and incomplete. Secondly, the pressure of procuring new business was occupying the very best people in systems engineers in sales efforts, often to the detriment of ongoing projects. Finally, the organization's capabilities for quickly specifying and testing a given design were inadequate, particularly given the tight time constraints which were usually present during a sales situation.

This last application area, that of providing design assistance during customer negotiations, was chosen as the subject for the prototype system we were to develop.

"CONCEPT DEVELOPMENT"

The technical support for negotiations required something which the client organization referred to as "concept development." This process was usually performed by a small team of individuals who were totally responsible for all sales presentations, technical design and proposal writing. Typically, a "concept development" effort would last for several months, during which time the proposed design, schedule and cost would be revised numerous times.

The negotiating team was normally comprised of at least three individuals:

- marketer;
- project manager; and
- systems engineer.

The marketer was responsible for all customer contacts prior to the time a contract was actually signed. The project manager was responsible for all schedules and costs. The systems engineer was responsible for devising an initial design and ensuring that the design would meet all of the customer's performance requirements.

The initial task faced by the systems engineer was often referred to as the "blank piece of paper" problem. That is, with little help from his colleagues and insufficient information from the customer, the engineer was charged with constructing a proposed design which would satisfy the customer's needs at a reasonable cost. The engineer's design constraints were conflicting: on the one hand, the design must be sufficiently simple so that it could be fully specified in a very short time period; on the other hand, the design must be sufficiently detailed to be convincing to the customer.

Each successive version of the preliminary design required a functional "walk-through" during which the customer was shown the precise behavior and logic of each component contained in the design, as well as the system's overall performance. This demonstration of the design's capabilities was often a long and difficult process due to the complexity of a typical ASRS design, and a great deal of expense (often non-recoverable) would be incurred in putting on convincing shows for a potential customer.

Frequent design modifications were almost always required during the negotiating period. These modifications were caused by a number of factors, including unsatisfactory design performance as well as customer-specified changes in the target functional requirements. In fact, our client's competitive position was often heavily dependent on the speed with which such modifications could be made and the results presented to the customer.

SUPPORT REQUIREMENTS

The overriding need of the negotiating team was the ability to conceive, specify and test a preliminary design very rapidly. It was stressed to us over and over again that the delay between the written conception and computerized testing of a particular design was often as much as two weeks. In addition to the competitive disadvantage caused by this delay, engineers were sometimes spending long, tedious hours simulating the behavior of their designs by hand.

Once initially specified, a given design usually went through extensive modification and retesting of the design, often on very short notice, during the negotiation process. These modifications were made to:

- governing parameters;
- component logic; and
- layout configuration.

Governing parameters included a variety of rates, capacities and statistical distributions which affected performance. Component logic included the details of how individual transactions were handled by specific pieces of equipment, as well as the methods whereby different parts of the plant communicated and interacted. Layout configuration, finally, was the description of how equipment was attached to form a network, both physically and electronically.

An additional type of support needed by the negotiating team was in convincing a potential customer that a proposed design met the customer's needs. Such a sales job usually consisted of a detailed, functional "walk-through" of the proposed design, and was often directed at management personnel from the customer's organization, who had little knowledge of ASRS principles. The combination of a high level of detail and a technically unsophisticated audience made this task extremely difficult. Moreover, it normally had to be repeated several times, once for each major set of modifications in the proposed design.

SYSTEM CONCEPT

Our initial objective was to apply AI technology in the support of the "concept development" process, not necessarily to build an expert system. The resulting system concept was, in fact, not an expert system, but rather an advanced decision-support tool that could be developed in a cost-effective way by the use of AI-based tools and techniques. Even though the ultimate objective of our client was to build a design-specific expert system, the development of a support tool was seen as the required first step in that development; this is discussed in more detail below in the section "Knowledge Engineering Strategy."

The prototype which we built came to be known as the "Graphical Interactive Simulation Tool" (GIST). We believed that this name conveyed precisely what the prototype was: a tool for experts to use, not an expert system for use by novices.

We envisioned that the primary user of our GIST prototype would be the senior systems engineer in charge of construction of a preliminary ASRS design. Others might occasionally use GIST, but we felt that it was critical to focus our efforts on the needs of the senior-level engineers whose skills were so badly needed elsewhere in the organization.

We felt strongly that the GIST prototype should be an integrated environment suited to evaluating a number of alternative ASRS designs within a short time period. In addition, GIST should provide a central repository for all design-related information during the negotiating process.

Another important aspect of the GIST prototype, we felt, was that it should represent explicitly the variety of physical objects present in an ASRS facility in a way which was intuitive and as close as possible to how the client's system engineers actually thought about those objects. In particular, we believed that the attributes and behaviors of those objects should be easily inspectable and modifiable by the user. The objects in question included:

- Components: for example, conveyor belts, cranes, racks, and diverters.
- Subsystems: small collections of components used as a logical unit, e.g., loop conveyors formed by a sequence of straight conveyor belts and diverting mechanisms.
- Transactions: the parts or products being stored and retrieved by the ASRS facility.

Moreover, as the name implies, it was imperative that the GIST prototype be as interactive as possible. More specifically, we believed that GIST should continuously provide useful information to the user while it was being operated.

Finally, we felt that it was not necessary for the GIST prototype to offer high-level design assistance. Rather, it was more important, at least initially, for the prototype to automatically take care of many of the details which made the job of configuring a particular design so time-consuming. Thus, the advice to be provided would be rather detailed, but would be at a low level of abstraction from a designer's point of view. Our client hoped and expected that high-level design expertise could be provided at a later time in the project.

KNOWLEDGE ENGINEERING STRATEGY

"Knowledge Engineering Strategy" refers to the series of steps required to capture those aspects of the design process in the GIST prototype. These steps included exploratory discussion, in-depth interviews concerning past design efforts and the post-programming efforts required to mold the prototype to the needs of the client's organization.

Four individuals from the client's staff participated in this process:

- two senior systems engineers;
- one marketer;
- one programmer.

All four individuals had worked together closely in the past on several different projects and were quite familiar with one another's working styles.

The participation of the programmer was necessary for two reasons:

- the prototype was largely to replace the simulator-related support heretofore provided by the programming staff; and
- this particular individual was to take over the development by the prototype after ADL's contract was completed.

A key aspect of our strategy was to study the experts' response to the evolving prototype. It has been our experience that sophisticated, AI-based tools almost always change the way experts approach their jobs. We felt that getting such a tool into the hands of the experts, and observing how they used it, would give us insights into their knowledge which was not obtainable in any other way.

Moreover, the ultimate aim of the client organization was to build an entire family of expert systems for providing high-quality design assistance throughout their organization. The decision-support facility provided by the GIST prototype, then, was seen as a first step towards that goal, since it was to contain a set of capabilities shared by all of those future expert systems.

IMPLEMENTATION STRATEGY

Our implementation strategy was straightforward. It was based on the premise that speed of prototype development was more important than concerns about how the fully-operational version of the prototype would be deployed in the field.

ADL was, therefore, able to take advantage of the superb programming environments available on various Lisp-based workstations. The earliest versions of the prototype were implemented in Interlisp on a Xerox 1108, later installed on a Symbolics 3600, after a conversion of the GIST prototype from Interlisp to Zetalisp, and finally demonstrated on a Texas Instrument Explorer.

Our primary software tool was IntelliCorp's Knowledge Engineering Environment (KEE). KEE was used for a number of reasons:

- it provided a robust rule interpreter;
- it has an excellent facility for object-oriented programming; and
- it allowed a transfer of the prototype from Interlisp to Zetalisp with a minimum of effort.

Due to significant overhead associated with KEE's user interface and error checking, the use of KEE increased the prototype's response time considerably, but not excessively. In fact, we planned on the eventual replacement of KEE with our own implementation of KEE's features in pure Zetalisp. Exploratory efforts showed that reprogramming of the prototype without KEE would increase the execution speed by a factor of one hundred or more. Nevertheless, the increased speed of prototype development due to using KEE warranted its use, and furthermore, the cost of using KEE was recovered in this phase of the project alone.

Special-purpose Lisp functions were used for the great bulk of the prototype's actual operation, including:

- control logic; and
- user interface.

In particular, Lisp was used to simulate supervisory control in an ASRS facility and to provide all algorithmic descriptions of component behavior. Virtually all of the input and output associated with the user interface were also programmed in Lisp, including menus, keyboard input and state displays.

OPERATIONAL FEATURES

As implemented, the GIST prototype has a number of features which make it an extremely useful tool for ASRS designers. We will now describe these features briefly before giving concrete examples of the system's output.

From a practical point of view, the most important feature of GIST is a general-purpose facility for defining an ASRS layout. With this facility, the user chooses components from a set of menus, locates them on the display using a pointing device, then connects them to make an ASRS component network. In addition, the user has the option of choosing often-used component subsystems (e.g., a loop conveyor) and logically treating that subsystem as a single component.

The most significant point about the layout facility is that graphical definition of the ASRS layout automatically results in a design which can be tested (i.e., simulated). In other words, there is a one-to-one correspondence between the graphical display and the underlying logic. Using object-based programming, symbolic and logical connections can be made equivalent.

A second feature of GIST which has proven useful is that, to the extent possible, component logic is driven by user-defined production rules which specify component behavior. For example, production rules can govern the choice of subsequent destinations when a transaction has been processed by any particular component. These production rules are inspectable and may be modified by a user with no programming experience whatsoever.

As a part of the user interface, GIST also has an extensive set of capabilities for providing a dynamic state display of a design operation. This includes simple animation of the design while it is being simulated, as well as numerous gauges and actuators which may be defined interactively by the user in a matter of seconds. In addition to providing the engineer with a visual display of how a particular design is performing, it is also an excellent sales tool for convincing a potential customer that a given design meets their performance requirements.

Finally, GIST has a wide array of flexible logic for automatically notifying the user when improper connections are made during the process of defining and testing a specific design. For example, there is a built-in set of constraints which specify exactly how component pairs may and may not be connected; these constraints define, loosely speaking, "plug compatibility" for ASRS components. Similarly, several types of bottlenecks are monitored during the testing of a design. When such bottlenecks occur, the user is notified and advised on how to take corrective action.

SYSTEM OPERATION

Now we would like to provide a few examples of how the user interacts with the GIST prototype during a typical session. In order to get a feel for how GIST is used, it is necessary to see the package actually being used, but simple screen images will suffice for our current purposes.

Figure 1 shows the basic interface in its initial state. Running down the left-hand side one can see the various menus through which the user indicates exactly what he or she wishes to do. The functionality of each menu is summarized below:

- "This Design"

  These are system-level commands for starting, stopping, loading and saving entire designs.

- "Connections"

  These allow the user to add or delete connections between components.

- "Available Clusters" and "Available Components"

  These menu items are chosen (or "moused") when the user wishes to add additional components to the design. "Clusters" refer to subsystems of components which can be treated logically as single components. "Components" are individual pieces of equipment. By mousing on a menu item, the user is presented with the current list of corresponding components (or clusters). For example, Figure 2 shows the list of choices presented when the user mouses on the item labelled "Decision Point." Each class of components (chute, conveyor, etc.) has its own icon which is used in all graphic layouts containing that component.

A simple design layout is shown in Figure 3. To gain an appreciation of how quickly such a layout can be specified, consider that the design shown in Figure 3 was put together in just over five minutes! Once a layout topology has been specified, the corresponding design can be simulated immediately, with no further modification.

Once a simulation has begun, the user is free to stop the simulation, make any modifications which seem reasonable (including rewriting rules, changing algorithms, etc.), and restart the simulation from exactly where it was stopped. Browsing facilities have also been included within GIST which allow the user to inspect the current state of any component or transaction within the system.

Figure 1

Initial State of User Interface

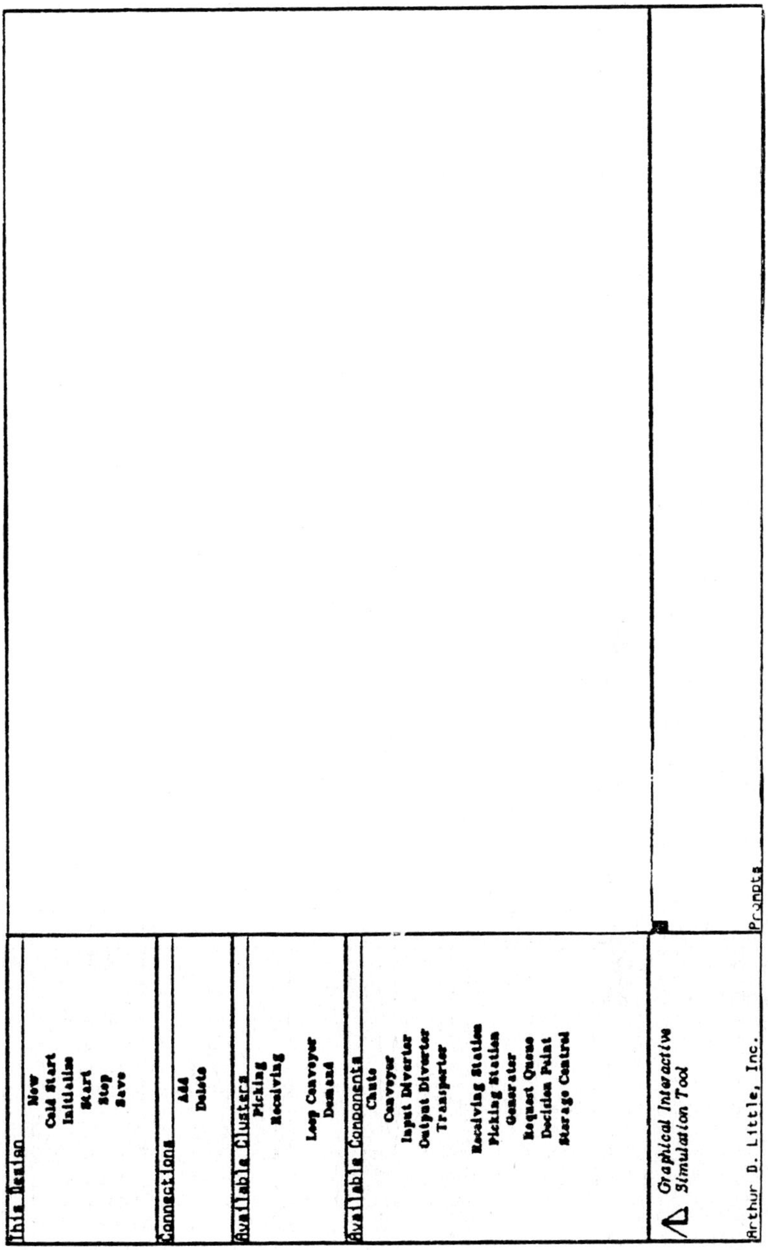

Figure 2

The User Chooses Menu Item "Decision Point"

(lower left)

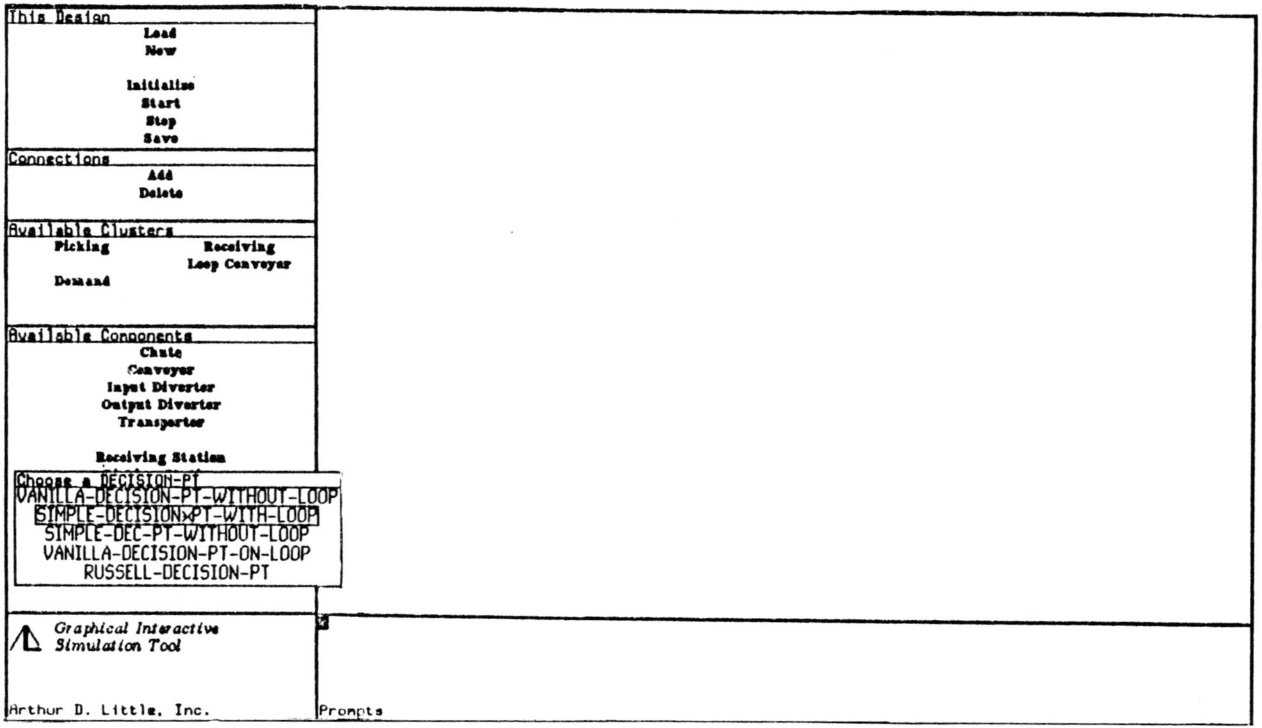

Figure 3

A Sample Layout

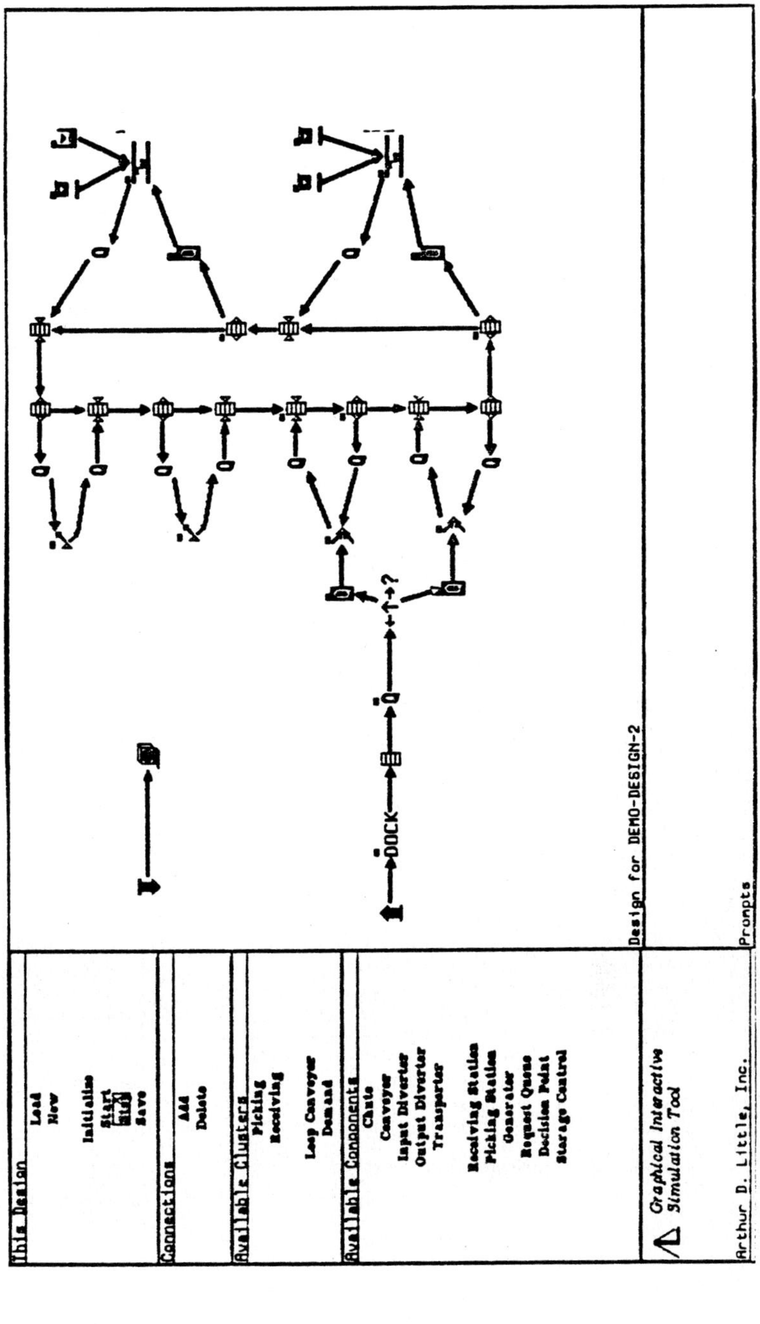

When the mouse is moved over some particular icon, and a mouse button is pressed, a menu will appear offering the user a set of options. One such option is "Modify Parameters" (not shown) which when moused, results in still another menu being presented which allows the user to change certain parameters which are often modified during the design process. Figure 4 shows the menu presented when the user chooses to modify those parameters associated with that type of component in which transactions are permanently stored, as well as the remainder of the display not shown in Figures 1-3.

Other aspects of the display as shown in Figure 4 are summarized below:

- Bottlenecks

  Icons shown in reverse-video show components held up by conditions at a downstream component.

- Gauges

  Several simple gauges are shown which are monitoring various statistics being gathered during the simulation of the illustrated design.

- Transactions

  The location of individual parts are shown by black bars at the upper-left corner of a component icon; the length of the bars is proportional to the number of transactions residing at that component.

Figure 5 shows a very crude example of the type of warning presented to the user when infeasibilities are found during the testing of a design (in this case, simply a warning that a capacity has been reached). Similar warnings are given when a component cannot handle a particular type of transaction or when some component has been idle for too long. The existing set of warnings may be removed or modified by the user, or new ones may be quickly defined as the need arises.

An additional feature not shown in the figures is the presentation of a warning (which may be overriden) when the user attempts to connect two components which are not "plug-compatible." These warnings attempt to prevent the user from putting together an unreasonable combination of components within a given design; for example, it is not reasonable to put a workstation at the end of a conveyor belt with no intervening buffer mechanism, even though it is physically possible. It is these sorts of simple errors which the "plug-compatible" warnings are designed to prevent.

Figure 4 Modifying A Component's Governing Parameters

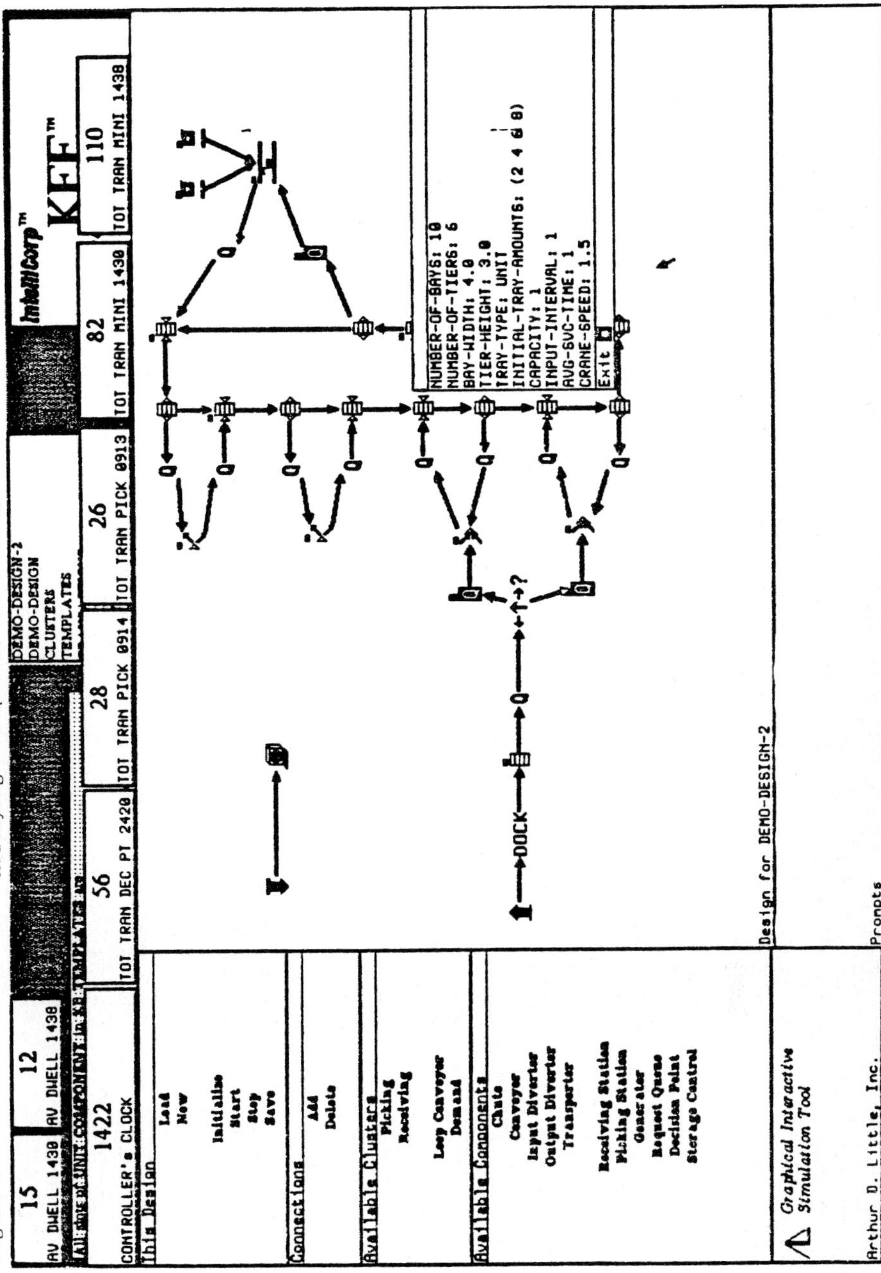

Figure 5 An Infeasibility Warning is Presented to the User

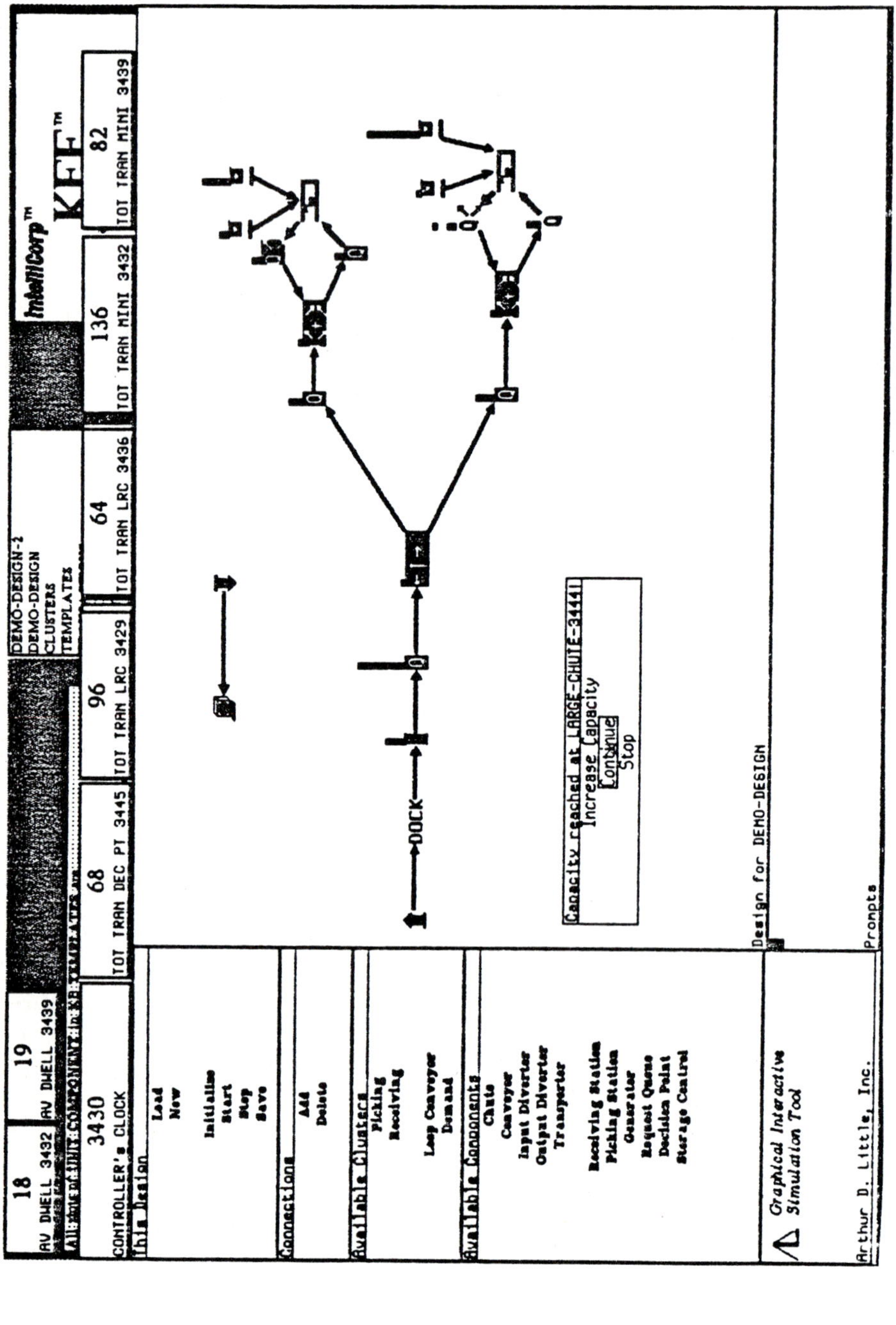

LESSONS LEARNED

During the development of the GIST prototype, a number of conclusions were drawn concerning the application of AI technology to design problems. The three major conclusions are as follows:

1. Interactive simulation is a "base technology" for design-related expert systems.
2. Knowledge-based simulation is an extremely powerful tool for complex design problems.
3. The addition of design expertise to a sophisticated tool (such as GIST) can and should be performed incrementally.

The ability to simulate a particular design interactively will be an integral part of virtually any expert system for design problems. In that sense, it may be considered a "base technology"--to use strategic planning terminology--in the same way that forging steel is a base technology for automobile manufacturing. The reason for this is very simple: design engineers are basically a skeptical group of individuals who must be convinced in concrete terms that a particular design will work before they trust the behavior of that design.

The capabilities of the GIST prototype are sufficient evidence for the conclusion that knowledge-based simulation is a powerful design tool. Using AI-based tools and techniques, GIST was constructed from scratch using less than one person-year of effort; conventional programming practices would have implied a development time of many times that amount. Furthermore, and more importantly, GIST shortened the amount of time to specify and test a given design from a few weeks to a few hours; this alone provides the justification for building the GIST prototype and points out the improvement in efficiency and flexibility which AI technology can offer.

Finally, the evolution of a tool like GIST must necessarily be a gradual one. As long as the proper amount of flexibility has been built into the initial design of the prototype, it is possible to incorporate large amounts of design-related expertise into a tool such as GIST before the fundamentals of the system architecture have to be modified. Moreover, we would argue that the incorporation of design expertise <u>should</u> be done incrementally, since the use of a sophisticated, GIST-like tool will eventually change the way that designing is performed, and in ways which are quite unpredictable at the outset of the prototype- development process.

ADDITIONAL USES OF AI TECHNOLOGY IN THE DESIGN PROCESS

Using our experience with the GIST prototype and its focus on ASRS design as a guide, we believe there are a variety of other uses for AI technology in the design process. To be more concrete and allow us to build on the preceding discussion, we will use the ASRS design process as an example and leave to the reader any extrapolations to other areas of design.

In our opinion, there are two principal types of design knowledge to be captured using AI-based techniques:

1. General design principles; and
2. Experience-based knowledge relating to ASRS operation.

These two types of knowledge are not only unique from a structural pint of view, they are also used at different stages of the design process and are obtained from different sources of expertise.

The capturing of expert-level design knowledge in a design support system would result in something resembling a "traditional" expert system. On the other hand, it is difficult to imagine such a system being used alone without a tool such as GIST to provide the capability of developing and evaluating a given design. The specific types of knowledge which might be captured in a design support system include:

- Conversion of customer needs to support requirements.

  In the initial stage of a design process, a great deal of time is spent translating what is said by a customer into testable performance specifications. There are simple, but highly useful, guides for how to perform this task.

- Selecting a preliminary layout.

  This corresponds to what was earlier referred to as the "blank piece of paper" problem. It would be by far the most difficult of all knowledge to capture, but could have correspondingly large benefits if done successfully. An intergrated library of generic designs would be useful at this point.

- Diagnosing a design.

  A non-optimum design often behaves such that it exhibits certain "signatures" which are familiar to the most expert designers. An expert system which focused on these problems would be very similar in principle to other diagnostic problems in medicine and finance.

- Taking corrective actions to improve a non-optimum design.

  This type of design support has been separated from the diagnostic problem mainly because it is one level of abstraction above the detection of design flaws. In particular, actions often involve conflicting design goals and corrective subjective choices concerning which goal(s) will dominate over others.

The capturing of experience-based knowledge relating to ASRS operations would be more difficult than capturing design principles. The primary reason for this difficulty is that the knowledge resides in numerous individuals throughout the organization, whereas design principles are known best by a relative handful of individuals. The precise form of experience-based knowledge tends to take the form of anecdotes which describe how particular operating problems were handled in the past, either by customers or by the engineering staff during the installation process. The concept here is that experience-based knowledge would be added to a design not as extra information directly affecting the design, but simply as a logically-connected set of notes to be referred to in the future when specific problems arose. We consider the following problem areas to be the most promising:

- Reconfiguration and reprogramming.

  There has been a great deal of publicity concerning programmable factories and warehouses. The reality is that it is extremely difficult to "reprogram" an entire facility when it comes time to, for example, change product lines or reconfigure a facility's layout. It was felt, by our client's staff, that AI might have something to offer in the way of obviating the need for extensive training if only they could encapsulate their own experiences and make them available to their customers on a permanent basis.

- Troubleshooting during start-up.

  The start-up of an ASRS facility is the time of greatest operational difficulty. This is largely because much of the finely-tuned software that controls an ASRS operation must be bypassed during the start-up process. There are standard "tricks" for such bypassing which could be embodied in an AI-based set of software and delivered as a standard part of an ASRS package.

- Responding to component failure(s).

  Any automated facility will have equipment failures which force the entire facility to be shut down unless there are either redundant pieces of equipment available or intelligent algorithms for bypassing the failed equipment while it is being repaired or replaced. According to our client's staff, there are standard "tricks" for getting around these problems similar in principle to those for start-up troubleshooting, which could be made a standard offering of any ASRS facility's control system.

CONCLUDING REMARKS

Our experience with the GIST prototype has shown that AI technology has a large potential for greatly increasing the efficiency and effectiveness of the design of ASRS facilities. We further believe that, with suitable modifications, many of the same ideas applied in the GIST prototype can be applied to other areas of design. We hope that we have been able to provide the reader with a glimpse into how AI tools and techniques can be applied to complex design problems, even when the outcome of that application is not necessarily the construction of an expert system. We also hope that we have indicated the amount of groundwork which is often required before a significant amount of expert knowledge can be captured in a set of software which is useful on a day-to-day basis.

As a final note, the client organization which sponsored this work is currently expanding on the concepts which were developed during the project. They have gone considerably beyond where ADL left off, and are now busily working on a new generation of software for use in their ASRS-related products. We wish them the best of luck.

ARTIFICIAL INTELLIGENCE IN MANUFACTURING
T. Bernold (editor)
Elsevier Science Publishers B.V. (North-Holland)

DESIGN, A CHALLENGE FOR AI

Jonckers Viviane

Artificial Intelligence Laboratory
Vrije Universiteit Brussel
Belgium

Design is an important activity both in economic terms and as an example of an interesting kind of human problem solving. Automating the design process will potentially improve cost and reliability of the design process and of the resulting artifact. Since Artificial Intelligence addresses the mechanisation of complex , knowledge intensive tasks it is a natural dicipline to turn to. Models of the design process need to be developed that can serve as foundations for the automatisation efforts. Requirements for the development of an intelligent Computer Aided Design system in the domain of VLSI are presented. It is discussed which requirements can be met using existing Artificial Intelligence technology and techniques. Some fundamental unsolved problems can be isolated as candidates for further research.

1. INTRODUCTION

Design is an important activity both in economic terms and as an example of an interesting kind of human problem solving. Automating the design process will potentially improve cost and reliability of the design process and of the resulting artifact. Since Artificial Intelligence addresses the mechanisation of complex , knowledge intensive tasks it is a natural dicipline to turn to.

There are two approaches to the relation between Artificial Intelligence and the automatisation of the design process. One is to view Artificial Intelligence as a technology that can provide and augment the automatisation and integration of some tasks or aspects of the design process. The second is to study the yet incompletely understood process of design and to develop better models of it that can become foundations for automatisation efforts. Both approaches challenge some fundamental research topics in Artificial Intelligence because of the diverse and complex nature of design.

Research in cooperation with Bell Telephone mfg. Co. and sponsered by the Belgian Institute for Scientific Research in Industry and Agriculture.

In the area of VLSI circuit design, automation is necessary if only to encompass the increasing complexity that arises from evolutions in the implementation technology. Requirements for the development of an intelligent Computer Aided Design system are presented. Some of these express criteria imposed by the future users of the system, while others emerge from the investigation of the nature of the design process. It is indicated which requirements can be met using existing Artificial Intelligence technology and techniques. Some fundamental unsolved problems will be isolated as candidates for further research.

2. DESIGN

2.1. The Nature of Design

The first intend of a design process is to obtain a description of a constructable artifact in a target medium that satisfies an initial formal or informal functional specification. Typical examples are the classic engineering diciplines including integrated circuit design where the goal is the construction of a chip performing a specified function in a target technology. Another well studied dicipline is software design where the goal is the development is a program (an artifact) performing a specified function on a target architecture or in a target programming language.

Several other requirements can be implicitly or explicitly present in a design situation :

- performance versus resource usage and cost of the resulting artifact
- ease and/or cost of manufacturing and testing the artifact
- documentation and reporting on the design process and on the result
- restrictions on length and cost of the design process itself

As thus design requires the integration of constraints imposed by the problem, the medium and the designer. This is reflexted in the exploratory nature of design. Constraints on the design come from many sources and usually there is no comprehensive theory that integrates constraints with design decisions. A designer can not immediately assess all consequences of design decisions and must explore different design possibilities tentatively.

This results in a evolutionary design process. The gap between specification and solution is usually so wide that a straightforward transformation from the first in the latter is impossible. The translation from initial specification to final solution involves iterations of synthesis and analysis or verification phases. Specifications are incrementally translated in structure and behaviour of the artifact. The evolving specifications must be tested and evaluated against the design goals and constraints. Problems with original solutions can occur, some impacts of design decision were not foreseen and are unwanted etc. This implies that changes are needed at different stages and levels of the design process. This adds again an extra level of complexity to the problem.

Design is recognised to be an expert activity in the sense that experienced designers are known to succeed better designs faster that novice designers. In most engineering diciplines there is an extensive theory of how the principles which underlie the operation of the artifact are applied in a particular design. In fact most of the knowledge thougth in engineering classes is a theory on how the artifacts work and how their behaviour can be derived from these fundamental principles. But skilled engineers know much more, they have a great deal of "tacit knowledge" which allows them to apply their knowledge efficiently to solve problems of design, synthesis and analysis. This tacit knowledge is not taugth explicitly in classes nor is it written down. It is usually considered informal and unteachable except by actual experience.

2.2. Design Methodology

A basic strategy for coping with complex problems is to find or impose structure on the problem which allows breaking it up into manageable pieces. The underlying idea is that a solution for the whole can be composed from the solution of the parts. This applies to the design of a complex artifacts in a complex field in two ways. The artifact to design is complex and can be obtained by designing parts which must be configured to make up the whole. This is translated in various shallow or strict structured or hierarchical design methodologies which exist in different domains. The requirements on the design process are so complex and involve so many aspects that it must be decomposed and stepwise covered. The designer must than cope with the interactions between the subproblems since they are seldom independent.

A radically different approach to overcome the complexity barrier of the design process is to lower that barrier. Although constraints are imposed on the design process from different angles it is in general still an underconstrained problem. The search space of possible solutions is to large. Additional constraints can be introduced on some aspects of the design process. For example imposing a style on the artifact reduces the complexity of the design process by restricting the space of solutions to be considered.

2.3. Computer Aided Design

Computer design aids began as programs performing the task of bookkeeping by storing design data in data bases. Routine tasks which involved the manipulation of large sets of data were next to be automated. As electronics technology evolved it became possible to replace paper and pencil with interactive graphical editors for design data input and graphical output for report and documentation facilities. The design is computer aided in the sense that the computer is used as a work environment with some nice features. It does not forget, it does not make mistakes in boring checklist tasks or data conversion tasks etc. This can be usefull for analysis and verification phases in the design. There is however very little support for the synthesis phases of design. This can easily be explained by the fact that only those tasks can be automated which are completely understood and as such can be formalised.

3. ARTIFICIAL INTELLIGENCE

In this section Artificial Intelligence tools, language and techniques are covered that relate to or are used in the development of complex systems.

3.1. AI Development Systems

A large part of Artificial Intelligence research addresses the automatisation of complex problems in diverse domains. Much of this research is so so advanced that is is exploratory in nature. It was early recognised that flexible and powerfull programming environments were needed to support this type of research. The most advanced interactive programming environments have been build in Artificial Intelligence centers. Typical examples are the LISP programming environments that exist on specialised LISP machines (Sandewall 1984 ; Greenblatt 1984). They give the researcher a well integrated set of tools to assist in experiment development. This technology is currently widespread and has been use as work environment in both research and development situations.

3.2. Expert Systems

Among the most significant successes of research in the field of Artificial Intelligence has been the development of powerfull computer systems known as expert or knowledge based systems. These systems are developed to represent and apply factual knowledge of specific domains of expertise to solve problems. Knowledge of the domain and knowledge underlying the human expertise is emphasised while the significance of domain independent problem solving methods apparently decrease. The ultimate goal is the construction of knowledge systems that autonomously apply expertise and equal or even surpass the performance of human experts.

Expert systems can be categorised into a few distinct types including interpretation systems, prediction systems, diagnosis systems, monitoring systems, configuration systems, planning systems and design systems (Hayes-Roth 1980). Experimental systems have been build for most of these types in different domains. Early successfull systems exist for application such as diagnosis and interpretation in both medical and technical applications. They have considerable impact on the domain and are followed by the development of various domain independent frameworks for building consultation systems. A considerable amount of effort for the construction of expert systems today goes in the process of knowledge acquisition. This is recognised to be the bottleneck for further successes in different domains.

3.3. Representation and problem solving

Building complex applications which involve different types of knowledge requires appropriate representation tools. Different language for representing that knowledge have been developed. The basic criteria for a knowledge representation language are expressive power, understandibility and accessibility. Experience has made it clear that a single formalism such as frames, predicate calculus, rules or constraints is not powerfull enough to satisfy all these criteria for all application domains. These difficulties helped motivate the development of various hybrid AI development systems. Besides providing structured representations for objects and classes in the application domain they incorporate an amount of inference to extent the explicitly encoded beliefs and facts. Additional facilities within the system to allow to describe domain dependent inference rules, decision rules, actions, simulation of object behaviour etc. Several of these languages are past the experimental stage and commercially available (Fikes 1985).

Investigating models of different types of reasoning to obtain systems with real problem solving capabilities keeps AI researchers busy since the early beginning. Different techniques emerged from these efforts ranging from general but weak methods to specific knowledge based methods.

3.4. Models for design

Efforts towards expert design systems are conducted mainly in the domains of software or algorithm design and integrated circuit design (Mostow 1985). The complex, exploratory and evolutionary nature of design implies that the following subjects must be considered in any automation effort of the design process.

- The state of the artifact under construction. This involves representations for the artifact at different intermediate stages of the design process, ranging from initial specification of functionality or behaviour to final specification of objects in the solution space. It requires the ability to represent the partial state of a design in which some but not all decisions have been made.

- The state of the design process. This is different from the description of the state of the artifact under construction. It situates the current relative stage of the design process such as which subproblems are worked out completely and which need further attention, which decisions have been made and which are pending.

- The design strategy. In most cases initial specification can not directly be transformed in a final solution. In complex systems the designer must factorise the design in subproblems. Interactions between the subproblems must than be coped with. This must be translated into a design strategy that will serve as foundation for the control of the design process.

- Decision rationales. Each time that different possibilities are offered by the design strategy one must be selected for exploration. This decision is based on rationales which depend on the different constraints that were imposed on the design and the design process. They come from many sources and must all be recognised and formalised to make them available for guiding the decision during the design process.

- The design history. The history of the design process must be explicit because it is the basis for reflection. This is needed because of the evolutionary nature of design. At each stage it can be necessary to reconsider earlier decisions. This is only possible when a history is present for each decision including justification and verification. This history can be helpfull for explanation and documentation facilities.

If all these aspects can be formalised, represented and integrated a usefull foundation for design experiments would be present.

4. VLSI DESIGN

4.1. Introduction

In the area of integrated circuit design the artifact is an electronic circuit and the target medium a specific technology. The initial specifications concern the functionality and characteristics of the desired circuit. The result consist in general of mask specifications for the designed circuit. In recent years progress in integrated circuit technology has lead to a vast increase in the complexity of the artifacts that can be constructed monolithically. On the opposite there are limitations to the complexity that an unaided human designer can control in any engineering situation. This results in a situation where the full capacity of VLSI technology can only be exploited when solutions are found to overcome the complexity barrier in the design process. An overview of the evolutions in electronic circuit design can be found in (Newton 1981).

To cope with the increasing complexity, design methodologies or specific design styles were introduced. Highly specialised design approaches such as standard cell and gate array faciliate the design process because they restrict the solution space to be considered. They don´t take advantage of the full capacity of the existing technology but can be economical interesting because of the lower design turn around time. The choice between implementing a circuit in gate array, standard cell or full custom approach can be viewed as the first decision that has to be taken.

For full custom VLSI, structured or hierarchical design methodology is introduced to control complexity by dividing a complex structure into more manageable substructures. The structure ensures that the circuits function and topology can be verified and that the resulting circuit can be tested. In addition it provides a reduction in design time by making the effective exploitation of both human and computer resources possible.

The class of circuit structures used as building blocks and the types of connections and decompositions allowed define the design style. In addition standard techniques for implementing commonly used, highly regular substructures such as PLA´s , memories and data-paths are considered. A well known example of structured VLSI design is introduced in (Mead & Conway 1980).

In general design methodologies divide the design process in different stages or levels:

- functional specification, a combination of formal and informal specifications of the circuits purpose, function, spatial and electric characteristics.

- system conception, the development of the general strategy or methodology to obtain the specified behaviour.

- system partitioning, the division of the global system in subsystems with special attention for dependency aspects, resulting in specifications for each subsystem.

- design of subsystems, each subsystem is designed from logic level through gate level and electric level to layout level.

- interconnection and integration of the designed subsystems resulting in a global circuit layout.

This does not imply that design is carried out strictly top-down. Several iterations through most levels are necessary to obtain a final reasonable and correct solution. Some verification tasks are claimed to be only possible in going bottom-up since some characteristics are influenced by the implementation of the lower levels. The final solution can be proved correct only after the chip is fabricated and tested.

At the same time efforts were made to automate certain aspects of the design task wich resulted in a variety of CAD tools. Systems for layout digitizations and interactive correction are extensively used by designers today. Design data is kept in data-bases, the input medium is graphical for some levels, lexical for others. Tasks that are relatively simple in nature but get complex and time consuming due to the quantity of data, were early candidates for automatisation. Typical examples are circuit analysis, simulation, design rule checking and mask generation. Another role for the computer that was early recognised was that of bookkeeper. Design data can be stored and combined with appropriate conversions be used as input for the tools mentioned above. Moreover providing access to the design data makes it an information soure and a communication link between designers. Nevertheless most CAD systems currently in use consist of a loose collection of programs requiring a large collection of data formats and often requiring manual intervention to move from one program to another.

4.2. Efforts in AI towards VLSI Design Systems

Designing digital circuits is an area of increasing interest and activity in AI research. Part of this interest comes from people and groups interested in developing their own special purpose circuits for AI applications. Others take the development of an intelligent design aid for VLSI as their researc topic while still another group view design as a source for challenging research on fundamental issues of human problem solving.

- The DPL/Daedalus design environment is an interactive VLSI design system implemented at MIT (Batali 1982). The system provides a procedural layout language and a graphic editor to build a hierarchical, object-oriented database. Special purpose procedures for building PLA´s and data-paths are included (Shrobe 1982).

- Palladio is a circuit design environment for experimenting with design methodologies and knowledge-based expert-design aids, under development at Stanford University (Brown 1983). It recognises the need of circuit designers for an integrated design environment that provides compatible tools covering the whole range of the design process. It includes means for explicitly representing, constructing and testing such design tools and languages. Several programming paradigms such as rule-based, data-oriented, object-oriented and logical reasoning are provided mainly by the experimental programming environments in which Palladio is implemented.

- Vexed is a knowledge based design assistant developed at Rutgers University (Mitchell 1984). It provides interactive aid to a designer implementing a circuit by suggesting and carrying out possible refinements on the specifications. A large part of its knowledge is composed of a set of implementation rules, each of which suggest some legal method for refining a function.

- The EURISKO system developed as an experiment in discovery (Lenat 1983) was applied to the problem of VLSI design. It succeeded in designing a revolutionary transistor assuming a three dimensional technology.

5. INTELLIGENT CAD SYSTEMS FOR VLSI

5.1. Specification

Traditional CAD systems provide the designer with an environment of usefull tools. Even when the environment is rich and completely integrated it remains passive. It is the designer who conducts the design process and he decides to make use of the available tools or not. This machine-aided human design paradigm can never succeed in covering all aspects of the design model employed by the designer. For example, if the system would provide a facility for maintaining the history of the design process it will require from the designer to input justifications for each decision he makes and to document each step he takes.

An intelligent CAD systems that guides the designer or designers through the design process and participates actively in it would be a first step towards autonomous machine design. Such systems need to be build on top of a complete integrated design environment. They augment the capabilities of traditional systems in different ways.

To obtain this functionality the system has a build in descriptive framework, reflexting all stages and aspects of the design model and methodology it is based on. This framework is used to control the selection of tasks to be worked on. The processing of these tasks is monitored and the coordination of the different subtasks is supported. Routine tasks or straightforward problems are worked on automatically by the system, other tasks are given to the designer with suggestions for tools and techniques to use on them.

In this framework it becomes possible to incorporate experiments on automating less obvious subtasks of the design process. The designer stays in control in the sense that he can refuse advise or reject solutions of subproblems proposed by the system. At any time the designer can take over control and proceed the design or parts of it without active cooperation of the system.

5.2. Requirements

This section presents some requirements for the development of an active, intelligent CAD system for VLSI design. There are a number of requirements that emerge from the need for a complete, flexible and integrated system that covers all stages of the design process. Other requirements are specific for the interactive nature of the proposed system. It is discussed which possibilities are offered by AI technology to meet these requirements and where further research will be necessary.

5.2.1. Representation

Covering all stages of the design process involves descriptions or representations for all aspects of the design process mentioned earlier. Design state must be explicit and design history must be maintained which requires appropriate representations. Developing powerfull and flexible representations for all these topics will be a key effort since it will affect the possibilities of the overal system. The design strategy and design rationales must be formalised in a framework that will assume control over the design process.

5.2.1.1. Design state and history.

Artifact descriptions at different abstraction levels, with appropriate detail for each abstraction level and at different stages of completion must be represented. Representations of the same part at different levels of abstraction must be related. Relations between subparts within one level of abstraction must be reflexted in the descriptions. The same part can at each level of abstraction require multiple representations or multiple views depending on the task which is manipulating, using or constructing the description.

Design history representations must provide easy access by both designer and system for inspection, maintenance and reflection. It will be the foundation for the explanation and justification facilities in the system.

Flexible frame-based languages with build in capacities for defining and exploiting class taxonomy, with support for multiple views or a context mechanism are suitable implementation mediums for this type of representation problems. One aspect needs additional investigation. The amount of data involved in a VLSI design is factors larger than the amount of data present in most applications build on frame based systems today. The amount of data can exceed the capacity or negatively affect the performance of these systems. Attention must be given to interfacing these systems with classic data-base technology to surpass these problems. This will also prove usefull for the development and maintenance of libraries of designed circuit or subcircuits.

5.2.1.2. Design strategy

The design strategy must be translated in a framework that supports the guidance of the designer through the design process. The framework reflects the top down refinement form initial specification to final solution throught he different stages and levels of the design process. Flexibility is a key issue since control may shift from the system to the designer and back at any time. Procedural descriptions of the steps to be taken does meet this need for flexibility. Goal oriented modelling of the design process can provide this functionality (Sacerdoti 75). The representations of goal-subgoal relationships in planning networks can be used as a basic descriptive framework.

Customised control structures for processing these goals must be installed. For each goal it must be evaluated if appropriate tools for automatic processing of the goal are available or not. The user must be advised on the different possibilities to proceed the design. Eventually the designer takes responsability for further processing of the current goal. The system will then resume a monitoring role and wait for opportunities or instructions to regain control.

This again asks for a major representation effort and requires a hybrid development systems that allows integrated representation of goal structure and explicit installation of control processes.

5.2.2. User-interfaces

The system is supposed to assist a human designer during the complete design process. This implies that interaction with the designer will be frequent and on different levels. This introduces the need for supportive, flexible user-interfaces which is a general requirement for interactive computer systems.

Each level of representation and each kind of task requires an appropriate interface. High quality graphics are needed for parts of the interface such as layout and diagram editors, specialised graphical or lexical interfaces are needed for input and inspection of data at multiple levels of representation. The interface should also adapt to different demands of different users of the sytem.

5.2.3. Development and Interfacing of Specialised Tools.

During the evolution of electronic circuit design numerous specific tools have been implemented. Integration of existing tools where appropriate will prove much more economical than rebuilding them. On the other hand the development of new or alternative tools that exploit the rich representations available in the system can be considered.

The system is supposed to suggest appropriate tools to the designer for different tasks. This requires knowledge in the system about available tools, their application domain, purpose and properties. The invocation of the tool on selected data and the interpretation and integration of the results can be automated when specifications on the input and output formats of the tools are explicitly available (Jonckers 85). This means that the system must incorporate some aspects of intelligent front ends as described in (Bundy 1984).

Much of the existing conventional techniques for simulation, verification, layout and routing tasks are computationally very expensive. Extrapolation of these techniques to the amount of data present in near future designs requires computer resources which are not available. Intelligent partitioning of the circuit solves this problem if the subproblems can be worked out independently and easily combined afterwards. Multiple level, hierarchical simulators are examples of this approach. Still some problems will remain, requiring a global approach. Therefore attention should be given to the development of approximately or qualitatively reasoning in these contexts following the work of (De Kleer 84).

Another solution lies in the elimination for the need of expensive computation where possible. Verification tasks can become unnecessary when the design methodology together with monitoring from the system ensures correctness by construction. For some tasks brute force methods can be replaced with more intelligent knowledge based solutions that exploit specific knowledge introduced for this purpose.

6. CONCLUSION

The nature of design makes the automatisation of the design process much harder than other kinds of expert problem solving. In the area of VLSI circuit design, automation of the design process is necessary to encompass the increasing complexity due to evolutions in the implementation technology. Some ideas for the development of an intelligent CAD system that guides the designer through all levels of the design process are presented. An explicit framework, reflexting the design strategy and methodology is introduced to assume control. Requirements on representation and integration and interface issues are discussed.

7. REFERENCES

Batali,J; Shrobe,H; Sussman,G; Weise,D (1982) The DPL/Daedalus Design Environment. In VLSI81 conference proceedings of the University of Edinburgh conference on VLSI.

Brown,H; et al (1983) Palladio: An Exploratory Environment for Circuit Design. in IEEE Computer December 83.

Bundy,A (1984) Intelligent Front Ends. In Bramer,M (eds) Research and Development in Expert Systems. Proceedings of the Fourth Technical Conference of the British Computer Society Specialist Group on Expert Systems.

De Kleer,J ; Bobrow,D (1984) Qualitative Reasoning with Higher Order Derivatives. In proceedings of the National Conference on Artificial Intelligence, Austin.

Greenblatt,D ; et al. (1984) The LISP Machine. In Barstow,R ; et al (eds) Integrated Programming Environments. Mc Graww Hill.

Hayes-Roth,F ; et al (eds) (1983) Building Expert Systems. Addison Wesley.

Jonckers,V (1984) Knowledge Based Selection and Coordination of Specialised Algorithms. In proceedings of the Fifth Internation Workshop on Expert Systems and their Applications, Avignon.

Lenat,D (1979) On Scientific Theory Formation : A case study using the AM program. In Machine Intelligence 9, Halstead Press, New-York.

Mead,C; Conway,L (1980) Introduction to VLSI. Addison Wesley, Readings Massachusetts.

Mitchell,T; Steinberg,L (1984) LEAP : A Learning Apprentice for VLSI Design. In proceedings of the Ninth International Joint Conference on Artificial Intelligence, Los Angeles.

Mostow,J (1985) Towards Better Models of the Design Process. In Artificial Intelligence, Volume VI, Number 1.

Newton,A (1981) Computer-Aided design of VLSI Circuits. In Proceddings of the IEEE, Volume 69, Number 10, october 1981.

Sandewall,E (1984) Programming in an Interactive Environment : The LISP Experience. In Barstow,R ; et al (eds) Integrated Programming Environments. Mc Graww Hill.

Sacerdoti,E (1975) The Nonlinear Nature of Plans. In proceedings of the Fourth International Joint Conference on Artificial Intelligence.

Shrobe,H (1982) The Data Path Generator. In proceedings of the conference on advanced research in VLSI, MIT.

ARTIFICIAL INTELLIGENCE IN MANUFACTURING
T. Bernold (editor)
Elsevier Science Publishers B.V. (North-Holland)

An Expert Program Generator Approach to Tailoring Standard Networking Products

C.M. Witkowski, P.J. Cornwell

Texas Instruments Industrial Controls Division

This paper describes TISPEC, a rule-based consulting system that assists with programming tailorable communication adapters between an Industrial Local Area Network (ILAN) and a wide variety of intelligent manufacturing equipment. A single hardware unit provides network connection to a class of equipment employing RS232, RS422 and RS423 serial interfaces. Several ILAN standards will eventually be addressed. These adapters perform all protocol conversion required between vendor-specific machine communication and a highly uniform set of network transactions. Each adapter is tailored to the requirements of a specific machine interface with a firmware module derived by TISPEC from a skeletal protocol model using the operator's response. TISPEC is embued with generalised knowledge relating to the types of machines for which adapters are prepared, and highly specific and specialised knowledge of the design of the adapter and the network. The latter is well partitioned, however, so that TISPEC may address other network standards. With this knowledge, TISPEC may pose questions of the user relating to details of the machine's operation, and the requirements of its communications interface. The result of a TISPEC session is a "configuration file" defining the customisation required for a new interface in a form suitable for direct installation into the network adapter. TISPEC is developed from our experiences in developing such configuration files manually. The primary value of this approach is in reducing the lead time of adapter production. although automated assistance in managing the potential of a large family of network adapters is important in supporting documentation and quality assurance.

INTRODUCTION

Given the proliferation of Expert Systems within the last five years, it is curious that so little attention has been paid to the value of these Artificial Intelligence Techniques to the area of manufacturing. The potential for improved through-put of work and cost cutting are slowing being demonstrated to this traditionally conservative market. Modern computing techniques are, however, penetrating this field; there can hardly be a page in a recent manufacturing trade journal that does not contain some reference to "Computer-Aided-this" or "Computer-Integrated-that". AI has still to play a major part in this explosion of interest in computers in manufacturing, but several important inroads have been made. Fox et al. (1982) have presented a method for scheduling within the Expert System paradigm. Gallanti et al. (1985) have designed a Knowledge-based approach to process control and Lalapet et al. (1985) have described an Expert System approach to materials selection for Computer Aided Design. Our present Expert Systems application is not created specifically for an end-user but is an internal tool to speed productisation of network interface units through the many phases that a new product must pass.

When in the body of the paper, the term "configuration" is discussed, it is used in a different sense from the RI/XCON team (McDermott, 1982; Polit, 1985). The initial scope of the TISPEC project was to produce a "program generator". During the course of the program the scope has been extended to encompass a number of other ancillary activities essential to the successful issue of an industrial software package. We shall see that "configuration" in this context is a multiplicity of operations.

The Domain and Problem Overivew

TISPEC serves to support the rapid development of software for Texas Instruments UNILINK (TM) Secondary Adapter, (note 1), recently announced as part of the Industrial Systems Division strategy of that corporation. UNILINK is a family of standard products which augment TI's existing range of Programmable Logic Controllers and Industrial Local Area Networking products. The specific purpose of the Adapter is to act as an interface between the OSI TIWAY1 network and other vendors' manufacturing equipment. The variety of machines and instruments for which Adapters are generated includes Industrial Robots, Computer Numerically Controlled (CNC) and Direct Numerical Control (DNC) machine tools; intelligent sensors and instruments; Programmable Logic Controllers (PLCs) and various computers, peripherals and display devices. The function of the Adapter unit is twofold. It performs a low-level protocol conversion between the packet-based HDLC network and any character-based serial interface on the attached machine. For equipment to be tractable to connection to the network in this manner, it must be fitted with a serial communication port designed as a minimum for attaching a local operator terminal, but the functionality to connect with a remote computer system is desirable. More significantly, the Adapter also performs a command level protocol conversion.

Each of the commands issued by the network host computer are transferred to the Adapter over the network. The Adapter translates the command in its uniform format to the form required by the specific machine, causing it to perform the function specified by the command. As the attached machine performs the command, it returns response messages which the Adapter analyses and translates into one of a restricted set of network response messages. The effect of the Adapter operation is to give all machines attached to the network a uniform interface as far as the application program is concerned.

The potential gains for such a strategy are numerous. The ability to connect multi-vendor machines with a regular multi-drop network alone, as opposed to individual long haul serial communications with a host computer, presents large reductions in cabling costs, immunity to electrical interference and computer communication overhead. The greatest advantages arise because, due to the protocol translation, each machine may be treated similarly by the network host application software. Additionally, the cost of supporting a number of special purpose and real-time software interfaces to a range of disparate machines in a manufacturing system will represent a sizable proportion of host compute resource utilisation. The Adapter allows a single uniform software package to drive the network. This in turn reduces system cost in terms of improved use of generic and portable control software which can be applied repeatably across differing installations. It also results in improved software maintainability.

The RS511 protocol defining application layer seven of General Motors Manufacturing Automation Protocol (MAP) is devised to address this problem. It is more ambitious than the Secondary Adapter, providing guidelines to equipment vendors regarding the design of communications interfaces to new machine controllers. The task of translating directly from existing machine protocols to a network implementing RS511 would be daunting indeed and seldom completely successful because of the non-uniform functionality many machines support at a communications interface. The Secondary Adapter Command Set is designed to account for the variability and vaguary of existing controller communications protocols. If required, each of the Secondary Adapter Commands does have a direct analogue within the RS511 description and a translation between the two representations is performed with a network-to-network Gateway. It is important to note that this provides an effective translation at all levels from media, via transport to application between a MAP environment and a machine.

Protocol translation is at best, problematic; the situation is compouinded by several factors. In the first place there is a very wide range of different equipment types to connect. Even within a similar group, for example the robots, the range of controller implementations and communication protocols accessing them, is enormous. This arises from a lack of past standardisation and guidelines, individual research on the part of vendors and product line related commercial reasons. In the second place, equipment from many such groups must be addressed by the Secondary Adapter with the same uniform approach if a generally useful solution is to be provided. Real manufacturing systems necessarily comprise disparate equipment selected for this particular benefits in the application. In the case of the Adapter approach the problem is further compounded by the usual constraints on the issue of a software product. The code only represents a small segment of the completed product. It must be designed, reviewed, coded, documented, tested, iterated and archived for manufacturing before release. The maintenance of current user documentation for a single software product in the industrial environment is itself a costly undertaking.

When applied to the potential range of communication products resulting from an approach such as the Secondary Adapter, these constraints rapidly make a standard product solution unsupportable without a high level of automated assistance.

TISPEC in Operation

TISPEC proceeds in several stages. The first is to determine whether the machine is suitable for connection to the TIWAYI network through a Secondary Adapter. This is decided principally on the availability of an RS-232 or similar communications port. In cases where no such communication is provided, an alternative to direct host control over the network can be suggested.

During the next stage, the type of machine is determined. Machines are broadly classified into robot (manipulators), machine tool (lathe, mill), instrument (probe, sensor or data acquisition device), computer peripheral (data entry terminal, printer, plotter, operator display), computer equipment (e.g., CAD/CAM workstation) or miscellaneous (not clearly falling into any one of these categories). Some types such as CNC machines with simple program transfer facilities and computer display devices are handled with a special portion of the rule set. They have highly stylised solutions to create the PIM.

The third stage determines the RS-232 (or equivalent) settings required for the successful communication between Adapter and attached machine. This in turn is dependent on the options available on the machine and the Adapter unit, and takes into account the potential data-rate mis-matches between the network, which has a finite bandwidth, and the attached machine, which may or may not support a full range of flow control options.

In the fourth stage, the form of the communications protocol is determined. We have determined a "skeletal protocol model" into which the majority of machines that we have investigated fall. A single command translation may be considered as having six major components:

preamble, command, acknowledgement, response-request, response, postamble.

It should be noted that not all machines support this full set: some, such as the preamble are often omitted. In many cases the various stages are truncated or combined. A number of machines use an industry standard transport protocol, such as DDCMP. The code to adequately cope with the complexity of such protocols must be called from a library of routines. An attempt is also made to determine whether each transaction with the machine follows a similar pattern. If so, the solution is made easier in that only command codes and parameters need be changed. A number of machines, particularly those designed for human, rather than computer use of the RS-232 link, do not follow a set pattern from command to command. The Expert System has little option but to assist the user through each step in a multi-stage command transaction. The Expert System helps by keeping track of various dialogues, depending on the success or failure of the requested activity. It also ensures that each possible outcome is considered within the dialogue transition table. Our experience has shown that a single command out/response back transaction over the network will often be translated into as many as ten transactions over the RS-232 link between attached machine and Adapter. Each path through a graph of connected machine transactions will often branch quite significantly according to success, or to the type of command failure.

In the fifth stage, each of the uniform network commands is mapped to the command or commands required by the attached machine to perform the specified activity. TISPEC ensures that each command is properly investigated and that appropriate information for the PIM user documentation is prepared. In many cases the information required is of a "pro-forma" nature. In every case, the user has the option to reject this and substitute or augment the section of user documentation with her or her own text. The same is true of the tests required for a complete Quality Control cycle. Many of the tests follow a pattern dependent on the choices made for each command. Once again the Expert System may elect to suppress a test or suggest further tests in any given instance. It is these differences of opinion between user and Expert System that form the primary input for changes to TISPEC.

Each command requires that special attention be paid to various differing factors. TISPEC both determines inconsistent responses relative to the requirements of the network and with respect to previous responses provided. TISPEC conducts a question and answer dialogue with its user, which is, if anything, rather verbose. The Expert System will ask for information in a variety of ways: simple yes/no questions, a menu of possible options, and input of numeric or alpha-numeric data are but a few. In most cases the system offers a default value, evaluated as the most appropriate in the current circumstances. At every stage a concise explanation is available to the user as to why information has been requested and what activity is in progress. Additional information may often be requested. This is in contrast to typical backward chaining approaches where the system provides information on demand only.

The output of an expert system session is a "Configuration File" which may be translated into the code for a PIM, and which provides all of the information necessary to tailor the user manual for that PIM. As a consequence of this, many of the Monolith rules in TISPEC contain considerable amounts of text in the right-hand-sides comprising the text of questions, parts of explanations, and segments of output file, as well as comments and housekeeping activities. It is a consequence of our forward chaining approach that the left-hand-sides of rules seldom contain a grate many conditions. To a greater extent, decisions are cumulative as opposed to a lot of questions followed by one large organising rule. We believe that in this and other respects the forward chaining approach offers many advantages over the backward chaining one, and that they both demonstrate considerable advantage over sequential programming in this class of problem.

Our Expert System approach to the "configuration" of UNILINK Secondary Adapter PIMS gives us a method for producing a rapid and effective solution to the difficult problem of protocol translation between control computers and a variety of intelligent manufacturing equipment. Demonstrations of automatically-assisted PIM generation show the considerable advantages that accrue when a (relatively) simple question and answer session can lead to a product in an advanced state of readiness, in terms of technical preparation, documentation and test plan, as well as reduced internal administration effort. We are presently investigating the use of extensions to TISPEC to perform diagnostic activities during testing and network simulation. Indeed, it is hard to see how a usefully wide range of PIMs could be generated without this support, given the level of staffing required, and the constraints imposed by the market.

Note 1: UNILINK is a registered trademark of Texas Instruments Incorporated.

References:

1. Fox M.S., Allen B. and Strohm G. (1982) "Job Shop Scheduling: An Investigation in Constraint-Directed Reasoning." Proc. AAAI-82 pp155-158.

2. Gallanti M., Guida G., Spampinato L. and Stefanini (1985) "Representing Procedural Knowledge in Expert Systems: an Application to Process Control." Proc. IJCAI-85 pp345-352.

3. Lalapet S., Zarefar H. and Wang B.P. (1985) "Role of an Expert System in the Materials Selection Phase of the Computer Aided Design of Gear Spurs." Proc. ROBEX-85, ISA pp57-61.

4. McDermott J. (1982) "RI: A Rule-based Configurer of Computer Systems." Artificial Intelligence Journal vol 19, pp39-88.

5. Polit S. (1985) :"RI and Beyond: AI Technology Transfer at DEC." The AI Magazine vol 5-4 (Winter 1985) pp76-78.

INTEGRATION

ARTIFICIAL INTELLIGENCE IN MANUFACTURING
T. Bernold (editor)
Elsevier Science Publishers B.V. (North-Holland)

THE ALVEY LARGE SCALE DEMONSTRATOR PROJECT "DESIGN TO PRODUCT"

Tim Smithers

Department of Artificial Intelligence
University of Edinburgh
Forrest Hill
Edinburgh EH1 2QL

The projects goals and the collaborating organizations taking part in it are introduced. By contrasting the conventional design and manufacturing methodology with the aims of the Design to Product project, the central concept behind the project is presented, as that of developing an integrated engineering design support system within an automated manufacturing facility. The Artificial Intelligence techniques which will be used to support the engineering design activity and its integration within an automated manufacturing and factory control system are described.

1.0 Introduction

The Alvey large scale demonstrator project "Design to Product" discussed in this paper is being managed by GEC Electrical Projects and involves nine collaborating sites. This paper presents an overall view of the project and the particular work being carried out by the Department of Artificial Intelligence, University of Edinburgh.

The need for improved efficiency and flexibility in the manufacturing industry, across all its activities, is now well recognized, particularly by the small batch manufacturing companies, and this need has stimulated the development of today's Computer Aided Design (CAD), Computer Aided Manufacture (CAM) and Computer Integrated Manufacturing (CIM) systems. The "Design to Product" project aims to develop by the application of Artificial Intelligence (AI) techniques and demonstrate in a realistic factory environment, the technology required for the next generation of computer integrated and flexible manufacturing systems. The project forms part of the programme of advanced collaborative research into Information Technology run by the Alvey Directorate, which was formed in 1983 following the recommendations of a committee chaired by John Alvey [1]. It is one of a small group of large scale demonstrator projects whose twin aims are to demonstrate the application of the new technologies in real environments and to provide exploitation led goals for the research programme [2].

The aim of the Design to Product project is to show that in the area of mechanical engineering the tools and techniques developed by Artificial Intelligence research can be applied to the integration and automation of the total product lifecycle: from design through manufacture, to field servicing and maintenance. The project, which started in April 1985, is split into two phases of two and a half years each. The first "pilot" phase is intended to prove the viability of the Design to Product concept. The second, "demonstrator" phase, will aim to establish a fully integrated demonstration of the principles developed during the pilot phase.

The consortium of universities and companies which are collaborating on the Design to Product project is made up of the Universities of Edinburgh, Leeds and Loughborough, together with GEC Electrical Projects, GEC Research (Marconi Research Centre), GEC Avionics, Lucas CAV Limited, and the National Engineering Laboratory (NEL). A more detailed list of the project collaborators and their respective areas of interest and expertise, is presented in Appendix I.

In order to put into context the concept behind the Design to Product project, I will briefly review the organization and inherent problems of present day engineering design and manufacturing methodology and the current developments in CAD/CAM and CIM systems. The way in which Design to Product proposes to solve these problems is then presented, followed by a description of the A.I. techniques which will be used. Finally a longer term view of the work of the project is discussed, together with an assessment of the time it will take before a Design-to-Product system will be working in a factory.

2.0 The Conventional Organization of Engineering Design and Manufacture

The procedures and operations carried out in the design and manufacture of engineered products are traditionally organized in a sequential way. Having identified the need and initial specification for a product the process starts with its conceptual, or schematic, design followed by its detailed design. There then follow separate stages for manufacturing planning, manufacture, and testing, before it is packaged up and sent out into the field, where it may be subject to regular maintenance and occasional repair procedures.

Throughout this sequential process knowledge about the particular product is being generated, some of which is then passed on for use in the next or subsequent stages. For example in the design of a round shaft element subjected to a torsional load which has a change in diameter part way along its length, from "D1" to "D2", and a shoulder fillet radius of "r", the values of D1, D2, and r will be related by a stress-concentration factor "K". In other words, for given values of D1, D2, and K r will have a specified value determined from a graphical representation of the relationship between D1, D2, r, and K, by the designer at design time. This knowledge about the size of the shoulder fillet radius will be passed on to the production engineer in the form of an engineering drawing of the component on which the dimensions D1, D2, and r will be specified. What will not be passed on, however, is the fact that the value of r depends on the value of D1 and D2 in a particular way, and in this example a possibly critical way. Thus, unless the production engineer, and the operator of the machine on which the component is turned, or the part programmer for the numerically controlled machine used, also know about the stress-concentration factor relating r to D1 and D2, there is nothing to prevent them from changing the value of r to one more convenient to the manufacturing plan: for example, to minimize tool changes.

When all the designing, manufacturing and testing operations are completed and the product is ready for dispatch, a great deal of knowledge about the particular product will have been generated; about its function, shape, manufacture and test performance. The problem is that some of this knowledge is lost between the different stages, and what survives is distributed throughout the organization. This loss and distribution of knowledge also means that some of it has to be recreated at different times and in different forms. This leads to inconsistencies and the inability to transfer knowledge in ways other than those allowed by the sequential organization of the different stages. This product knowledge generation, transfer, loss and regeneration is illustrated in figure 1 (see overleaf), where product knowledge generated is represented by the vertical axis and the sequential design and manufacturing stages are represented by the horizontal axis. The generation of product knowledge is then described by a rising "saw-tooth" form as knowledge is generated during each stage and some of it is lost between stages.

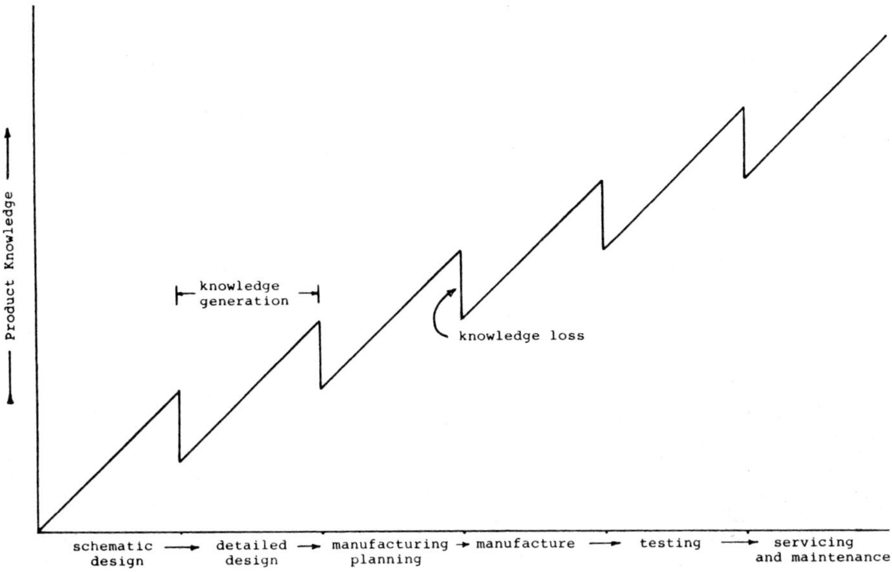

Figure 1. - Product generation and loss between sequential stages

2.1 CAD/CAM and its current Developments

The introduction of computer based techniques into the traditional engineering design and manufacturing procedures, in the form of computer aided drafting/design and computer aided manufacturing systems, has not changed the fundamentally sequential process described in 2.0: what they have done is to improve the efficiency and speed with which certain stages in the sequence are performed. By their support of standard parts, and part program libraries, they have improved the knowledge distribution problem in certain areas, but because they still support the sequential knowledge transfer concept they do not do much to improve the knowledge loss and regeneration problem. The introduction of new technologies into this sequential and departmentalized approach has thus lead to the creation of "islands of automation" [3].

The present developments in the CAD/CAM and CIM areas can be seen as attempts to link existing islands of computer based aids or automation: computer aided drafting to NC cutter path generation, and NC cutter path generation to NC part program preparation and planning, and part programming to production control, for example. However, these islands are still part of the traditional sequence of the design and manufacturing stages, and joining them up with common data bases and computer controlled data flow, or knowledge flow, does not remove the inefficiencies inherent within this type of organization. In other words, although perhaps smoothed in places, the rising "saw-tooth" product knowledge form remains. The feedforward and feedback problems that occur in the sequential organization of engineering design and manufacture are also not removed, nor are the data consistency and management problems. This failure to achieve real integration using current CAD/CAM and CIM systems means that the integrated inter-disciplinary functioning and continuity of operations which is required to improve the efficiency and flexibility of a manufacturing company is not realized [4].

2.2 The Concept of Design to Product

The aim of the Design to Product project is, therefore, to remove the losses in the product knowledge generation cycle, leaving a smoothly rising line, and to solve the knowledge consistency and management problems inherent in existing design and manufacturing methodologies. The concept with which it is proposed these aims will be met is that of an integrated and unified product description, whose consistency will be maintained and managed by the Design to Product system, and which will contain all the knowledge generated about a particular product at any time. At the heart of the Design to Product System lies the Designer System being developed at Edinburgh.

Thus, it is the commonality of the knowledge required and generated during a product´s design, manufacture and serviceable life which is seen as the underlying unifying and integrating factor, not the computer based nature of some of the different stages which occur during design and manufacture, as in the philosophy behind current CAD/CAM and CIM developments. The Designer System has therefore to be seen, not as one end of a sequence of design, planning and manufacturing stages, but at the centre of the whole product generation activity. This central position occupied by the Designer System means the introduction of flexibility into the traditional sequential ordering of the various stages, thus more easily allowing aspects within them to influence aspects within other stages. In other words, the product design activity is placed at the centre of the product generation process where each stage can more easily and directly influence the others in an integrated and consistent manner.

3.0 Artificial Intelligence Techniques

The concept of a unified and integrated knowledge base at the heart of the design and manufacturing activity provides an ideal application for some of the techniques which have come out of Artificial Intelligence research. The support of the various design activities is also seen as an important application of AI research, and of particular importance to the Design to Product project is the work done in the areas of knowledge representation, algebraic equation solving, geometric and spatial relationship reasoning, and automatic robot and manufacturing planning, design support and verification.

The way in which these AI techniques will be used and further developed is not yet completely defined as the project is only in its early stages, but the description of the current form of the Designer System and its various components follows, together with the identification of other areas within the Design to Product system in which AI techniques might be applied.

3.1 The Designer System

The purpose of the Designer System is to assist in the activity of mechanical engineering design and manufacturing planning, which is seen as being a part of the design process. The Designer System being developed at Edinburgh is a system for exploring the space of possible designs via the consideration of different design assumptions, and providing the means to assess and analyze the suitability of a particular design, or set of designs. It will also provide the mechanism for maintaining a consistent representation of the design, or designs, being worked on.

Within what is called the Designer System there are a number of sub-systems which provide the different types of functionality needed to support engineering design. They come under the broad headings of:

i) knowledge representation,
ii) design space exploration and design description,
iii) inference engines,
iv) specialists, and,
v) user interface,

and are presented in more detail in the following sections.

3.2 The Functional Unit Module Taxonomy

The formalism being constructed to represent the engineering knowledge required by the Designer System is described as the Functional Unit Module Taxonomy. It is a hierarchical organization of engineering entities which have a concrete referent; the term module being borrowed from the work of Barrow [5] on digital hardware design. Each module can be thought of as a package of knowledge about a particular engineering functional unit, different types of representation being used for the different types of knowledge contained within a module. Examples of the engineering entities for which there might be modules are energy converters, motors, shafts, keyways, bearings, oilways and tapped holes. An illustration of part of the module taxonomy for key joints is given in figure 2.

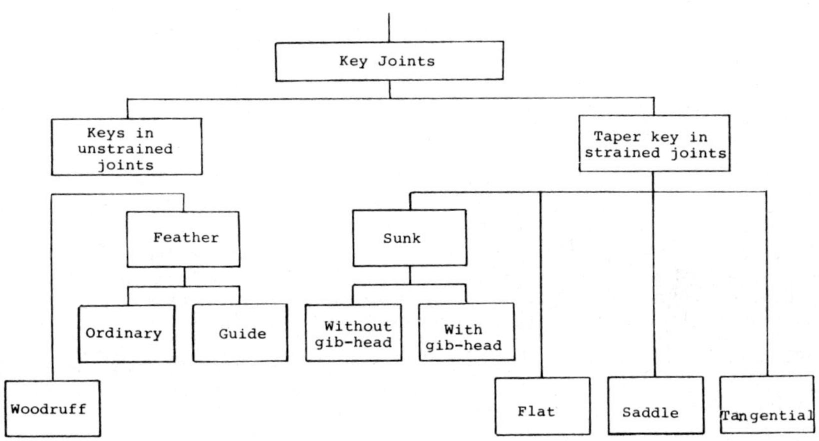

Figure 2. - Part of what the Functional Unit Module Taxonomy might look like for Key Joints

Using the Designer System a design is specified in terms of modules and connections between ports of these modules. Ports serve to define the standard ways in which one module may interact with another. A DC motor, for example, will have an input port where the current enters, an output port, where the torque goes out, and a support port, which provides the mechanical support for the motor. These connections define a relational structure on the design. Modules have parameters, which are symbols denoting quantities which are determined during design, and variables, which are symbols denoting quantities which vary during the operation of the module. For example the DC motor has as one of its parameters its maximum power output, and as one of its variables the rotational speed of its shaft.

At the lowest level of the Functional Unit Module Taxonomy the modules will be actual engineering components. Most of the knowledge for these modules will therefore be that presently provided in manufacturers component catalogues. The taxonomy thus also provides the means by which standardization in the use of certain types of components can be introduced into the design and manufacturing system.

3.3 The Assumption Based Truth Maintenance System

The activity of design involves the exploration of the space of possible designs. The exploratory nature of design is treated by regarding any statement made by the designer to the system as an assumption, which may or may not form part of the final design. For example, a designer may wish to consider the possibility of using an electric motor or a hydraulic motor in a given product, or at a more detailed level, the possibility of using a journal bearing versus a rolling element bearing. While some of the design work associated with a particular assumption may be lost if an alternative assumption is finally chosen, it is important to avoid losing more than is necessary since some of it may be used again when making further assumptions. This problem of maintaining "true" knowledge in a system where facts are being continually changed and added is one frequently faced in AI research, and there have been a number of approaches developed to deal with it. The system being developed for the Designer System to deal with the assumptions made by designers during the design of a product is called an Assumption Based Truth Maintenance System (ABTMS), and is based on the work of deKleer [6]. As well as assumptions about the existence of particular modules, the designer will also make assumptions about the values of module parameters, the connections between module ports, the spatial relationships between their features, and the grouping of modules to form parts.

A number of "engines" within the Designer System will be used to derive, or infer, consequences of the assumptions made by the designer. They will work in a forward chaining mode, although it will be necessary to restrict their operation in order to avoid an undesirable proliferation of consequences.

3.4 Design Description Document

The term Design Description Document is used to refer to the body of knowledge about a product which is built up using the Designer System and other sub-systems of the Design to Product system: it is where the description of a design, or set of designs, is maintained in terms of functional unit modules.

The process of design is thus the construction of a design description, or design descriptions, within the Design Description Document in terms of a set of assumptions and consequences. Each assumption corresponds to a design decision. There is no requirement for the assumptions to be consistent. For example the assumption that the parameter "p" of a module has the value 10 units might coexist with the assumption that p has the value 20 units. However a final design is characterized by a set of assumptions which must not be inconsistent. So the two values for p quoted above can only form part of alternative possible designs.

While a small design exercise could be carried out with no additional structure on the ABTMS, serious design will require the provision of a means to focus the design space exploration by setting a limit on the set of assumptions and consequences used to infer new consequences, thus preventing the generation of possible consequences that are not of immediate concern to the design. For example, in the case of a gear box design, the initial focus will be on the gear teeth, followed by the location of the bearings on the shafts, the loading of these bearings, the analysis of the shafts for stiffness and strength, the detailing of bearings, housings and gear wheels, etc.

Conceptually the Design Description Document is at the centre of the Design to Product system where it holds the knowledge about a product generated and required during its design, manufacture, and service life.

3.5 The Algebraic Manipulation and Solution Engine

The Algebraic Manipulation and Solution Engine, which is presently being based on the PRESS system developed by Sterling et al [7] will be used to support the symbolic equation manipulation and solution activities required by engineering design. It will also be used to follow paths in the relational structure of the design mediated by constraint equations, or other types of symbolic representations, and to infer new relationships.

3.6 The Spatial Relationship Reasoning Engine

This engine will be based on the RAPT inference engine, which forms part of the RAPT Robot Programming System developed by Popplestone et al [8],[9], and will be used for inferring module locations from defined spatial relationships. This engine will usually work in association with the Geometric Modelling engine.

3.7 The Geometric Modelling Engine

The representation of shape is supported by a Constructive Solid Geometry (CSG) modeller [10], which is being developed by Leeds University, building on experience gained with their Noname system [11]. CSG modellers represent shapes as a boolean combination of relocated primitive shapes, often referred to as CSG trees. The use of a solid geometry modeller is required so that space occupancy and intersection questions can be answered, something which is not possible with wire-frame modellers. The modeller engine thus provides the three dimensional geometry and spatial occupancy reasoning capability required to support mechanical engineering design.

3.8 Specialists

Apart from the support for designing the functional aspects of mechanical parts, there is also a need to support the "design for" type of activity; design for assembly, design for machining, design for maintenance, design for test, and design for cost, for example. These specialized aspects of design will be supported partly by including relevant knowledge in the functional unit modules of the taxonomy, and partly by rule based sub-systems which are refered to as Specialists. The two Specialists which are receiving particular attention at the moment are for supporting design for assembly, and design for machining.

3.9 The User Interface

To build a user interface for the Designer System is a large design problem in itself. One thing is clear, however, and that is that the continuing development of large high resolution screens supporting multi-windowing and non-keyboard interaction devices will be of particular importance. There has been, and continues to be, a good deal of work on how man-machine interfaces should look and behave, and the HUSAT research centre at Loughborough is a collaborator in the Design to Product project for their expertise in this area. A particular problem faced when designing a new tool as powerful as the Designer System, is that you cannot easily get a true picture of how engineers will use it before it is built. One approach to this problem being developed is called "user modelling" and involves the interface to the system maintaining a "model", or "models" of the people who use the system, so that it can adjust its response to a particular user, or type of user.

4.0 Manufacturing Planning and Scheduling

In the traditional sequential and departmentalized organization of engineering design and manufacture, planning and scheduling is often restricted to a slot between design and manufacture. In reality, however, the need to plan and schedule a product´s manufacture starts from the moment designing begins, and finishes when the last manufacturing, or testing, operation has been completed. It therefore needs to be supported at all stages of the product´s design and manufacture. In the Design to Product system, which places the design activity at the centre of the product generation system, where it is surrounded by, and interacts with, all the other components of the factory system, the scope of planning and scheduling can, with benefit, be extended to a much more diverse and varied set of problems. Within the Design to Product project it will only be possible to consider a sub-set of these problems.

5.0 Automated Manufacture and Factory Control

Automated manufacture includes both the use of machine tools and robots and their control within a manufacturing environment. If it is to be automated, it is an environment for which techniques to deal with the inherent uncertainties need to be developed. Within the Design to Product project work will be carried out by the GEC Research Marconi Research Centre and the GEC Electrical Projects Factory Automation Systems Technology (FAST) division on the programming and control of sensor based robot assembly cells and machining cells. This will include numerically controlled machine tools and measuring machines, and an automatic guided vehicle system for transporting materials and parts around the factory. The Design Description Document together with a dynamic scheduler will provide the control information for the manufacturing and transportation systems

6.0 Field Service and Maintenance

During its service life a product will need documentation for installation, operational adjustment and setting, and routine maintenance and fault correction. This information will be derived from the Design Description Document either automatically, as printed documents, or as a knowledge base supporting a portable terminal. Work in this area will be carried out by GEC Avionics during the project.

7.0 Implementation

The Edinburgh part of the Design to Product project is being implemented in the POPLOG programming system [12] running under the UNIX operating system on a network of Sun Microsystem computers. The POPLOG programming system provides three programming languages; POP-11, PROLOG [13], [14], and LISP, which can be freely mixed. It also provides an integrated screen editor which allows incremental compilation of code modules into existing compiled code.

In terms of hardware it is already clear that before a Design to Product system can realistically be implemented in a factory environment new and more powerful hardware systems will be required. Of particular importance is the current research into relational database hardware and parallel processor architectures for declarative languages.

8.0 Discussion

8.1 Modularity of Design

The use of functional unit modules to describe a design, or set of designs, should not be seen as implying "modular design", since the modules referred to in this paper are basic engineering functional entities. In order that the Designer System can contain definitions of complex higher level modules, and not coerce a modular form of design, the concept of modular identification has to be introduced. This allows the designer to decide that two modules which were originally considered to be distinct are in fact identical. For example an electric motor always has a frame, which supports its working parts, and a transmission likewise needs a frame. In an optimized design the frame of the motor and transmission may be identified as being the same.

8.2 Knowledge Encoding

The amount of knowledge to be encoded about specific engineering entities is large, and the use of Prolog clauses, although suitable from the systems point of view, is cumbersome from the encoders point of view. Thus the taxonomy will be encoded in an external form and automatically translated into the internal Prolog form.

9.0 A Long Term View for Design to Product

There are two things to say about the long term view for Design to Product. First, having described the aim of the project as being the development of the technology for the next generation of CIM systems, it should be noted that new generations take about ten years to grow up. Thus, bearing in mind that the application of Artificial Intelligence techniques on the scale of Design to Product has not before been attempted, and that its realistic implementation will also depend upon the results of the current specialized hardware research, the day that a complete Design to Product system is working in a factory is probably further than five years away. The application of parts, or sub-sets, of the system being developed by the Design to Product project should, however, be possible before then.

Secondly, the accepted wisdom about the organization of manufacturing industry is that the various activities involved should revolve around the financial control and management of the company, or companies. The need for change in this conventional organization is increasingly being recognized by the people and organizations involved. The opportunity for redesign, and reorganization, offered by the transition to new technology is one that cannot be ignored since it provides the only opportunity for an effective, and relatively painless change. The universal and integrated design and manufacturing knowledge generation and maintenance capability, provided by the Design to Product concept, allows design to have its properly central position at last. The implementation of the Design to Product concept, as presented in this paper, is seen as depending to a large extent on the successful application of Artificial Intelligence techniques.

10.0 Acknowledgements

The work being carried out at the Department of Artificial Intelligence, Edinburgh University, on the Design to Product project comes under the Alvey programme of research and is funded by H.M. Government, through the Science and Engineering Research Council. Additional funding for other parts of the project is being provided by GEC plc, Lucas CAV Limited, and the Department of Trade and Industry. I express my thanks to my colleagues in the Edinburgh Design to Product team: Robin Popplestone (principal investigator), Jonathan Corney, Gideon Sahar, Karl Millington, Anastasia Koutsou, Sue Renton, and Anne Elder, and to Chris Malcolm of the Department's Robotics group, for their help and advice while writing this paper.

11.0 References

[1] A Programme for Advanced Information Technology, the Report of the Alvey Committee, Department of Trade and Industry, HM Stationary Office, 1982.

[2] Alvey Programme Annual Report 1984. published by The Institution of Electrical Engineers, November 1984.

[3] Mather, D.V., Computer Integrated Manufacture, International Conference on The Development of Flexible Automation Systems, July 1984. Institution of Electrical Engineers conference publication no. 237.

[4] Jefferies, R.S., The Integrated Engineer and his role in Flexible Automation Systems, International Conference on The Development of Flexible Automation Systems, July 1984. Institution of Electrical Engineers conference publication no. 237.

[5] Barrow, H.G., Proving the Correctness of Digital Hardware Designs, Proceeding of the National Conference on Artificial Intelligence, Washington, DC, 1983.

[6] deKleer, J., Choices without backtracking, Proceedings of the Conference of the American Association for Artificial Intelligence, 1984.

[7] Sterling, L., Bundy, A., Byrd, L., O'Keefe, R., and Silver, B., Solving Symbolic Equations with PRESS, DAI Research Paper No. 171, University of Edinburgh, Department of Artificial Intelligence.

[8] Popplestone, R.J., Ambler, A.P., and Bellos, I., An Interpreter for a Language for Describing Assemblies, Artificial Intelligence Vol 14 No 1, pp 79-107, 1980.

[9] Corner, D.F., Ambler, A.P., and Popplestone, R.J., Reasoning about the Spatial Relationships Derived from a RAPT Program for Describing Assembly by Robot, Proceedings of the 8th International Joint Conference on Artificial Intelligence, Karlsruhe, DBR, 1983.

[10] Requicha, A.A.G. and Tilove, R.B., Mathematical Foundations of Constructive Solid Geometry: General Topology of Closed Regular Sets, TM27a, Production Automation Project, University of Rochester, USA, 1978.

[11] Armstrong et al, Noname Description and Users Manual, Department of Mechanical Engineering, Leeds University, 1982.

[12] Hardy, S., A New Software Environment for List-Processing and Logic Programming, in Artificial Intelligence, Tools, Techniques, and Applications, eds O´Shea and Eisenstadt, Harper and Row N.Y., 1984.

[13] Barrett, R., Ramsay, A., and Sloman, A., POP-11, A Practical Language for Artificial Intelligence, Ellis Horwood, 1985.

[14] Clocksin, W. and Mellish, C., Programming in Prolog, Springer Verlag N.Y., 1981.

Appendix I - Project Collaborators

The consortium of companies and universities collaborating in the Design to Product project, and their particular areas of interest is as follows:

1) GEC Electrical Projects, Rugby, Systems Design division - Overall project management.

2) GEC Electrical Projects, Rugby, Factory Automated Systems Technology (FAST) division - DNC machining cell and factory control and automation.

3) GEC Research, Marconi Research Centre, Great Baddow - Flexible sensor-based robotic assembly.

4) GEC Avionics, Rochester - Advanced servicing and maintenance facilities.

5) Lucas CAV Limited, Gillingham - System users and final demonstration site.

6) National Engineering Laboratory, East Kilbride - Robot calibration.

7) Edinburgh University, Department of Artificial Intelligence - The application of A.I. techniques to the development of an engineering design support system.

8) Leeds University, Department of Mechanical Engineering - Geometric solid modelling and machine tool path generation.

9) Loughborough University, Department of Engineering Production - Automatic machine tool instruction.

10) Loughborough University, Human Sciences in Advanced Technology Research Centre (HUSAT) - Man-machine interfaces.

ARTIFICIAL INTELLIGENCE IN MANUFACTURING
T. Bernold (editor)
Elsevier Science Publishers B.V. (North-Holland)

THE METHODOLOGY AND ARCHITECTURE OF COMPUTER INTEGRATED MANUFACTURING

by

E.G.R. GERELLE

and

PH.G. KIRMSER

ABSTRACT

A general framework for Computer Integrated Manufacturing is proposed. It consists of a methodology for achieving the objectives of CIM and an architecture for the technology of CIM. The position of Artificial Intelligence is then presented in terms of this framework.

INTRODUCTION

Manufacturing companies are under increasing pressure from the economic environment, in particular from markets, governments, employees and their competition. For a company to succeed, it is becoming essential that its objectives concerning the choice of products, the design and manufacturing processes, and the relations with customers and suppliers be clearly defined.
CIM is a corporate philosophy which supports the business goals of a company by its integrated approach to information and the use of technology. Computer Integrated Manufacturing represents a revolutionaly approach to achieving corporate objectives in a world that is increasingly dependent on information and on computer technology. CIM is applicable to both small companies and large corporations, to manufacturers of all types of products, from food and chemicals to computers and automobiles.

METHODOLOGY

Many different problems must be solved in order to achieve the corporate objectives and the benefits expected in implementing CIM. Different techniques must be applied by different parts of the organization, and serious communications problems may arise if there is no consistent approach to attack the problems. The CIM methodology provides a unified approach to achieving the objectives of CIM. The methodology has three basic phases: modelling (defining problems and building models), description (finding and evaluating possible solutions), and control (implementing solutions).

The methodology is problem and technique independent and can be used across the organisation, thus facilitating communication and integration between different departments. Due to the company wide nature of CIM it is important that an overall CIM strategy be defined well before any attempts to select or implement technology are made. The CIM strategy statement must include an assessment of the current situation, a vision of the future CIM environment and the process to link the two. The CIM methodology is illustrated by applying it to the development of a CIM strategy for a company.

The objectives of the company should first be defined. A useful technique is to build high level maps describing the relationship between the company and its environment (business model) and the internal relationships between departments (enterprise model). Setting up a glossary of terms with definitions from different parts of the organisation is also useful, serving to promote communication and understanding of different perspectives. Techniques for modelling the enterprise and its business environment include structured analysis and forecasting methods. These can be both informal diagrams, or simulation and mathematical equations, depending on the nature of the problem being analysed.

Once models of the enterprise and its environment have been developed, the behaviour of these models must be understood. The behaviour can be obtained by solving the equations, probing the simulator, or by codifying past experience in a set of ad hoc rules. The behaviour is classified as processes (behaviour of the enterprise), constraints (limits on the processes) and metrics (functions to be optimised). The processes are represented by parameters which describe the inputs and outputs of the enterprise. These include actual production rates, storage and process times. The constraints limit the processes to those which are allowable. The metrics are used to support decisions concerning the selection of the "best" allowable processes. Metrics evaluate performance in terms of productivity, quality, time to market, etc

The third phase addresses the business strategy, i.e. operational decisions used to run the enterprise and control the actual processes that have been selected. The relationship of processes, constraints, and metrics not only determines the current operation of the enterprise, but also identifies the critical areas where automation is required, i.e. the Automation Strategy. It is usual to have some conflict between the constraints and metrics. In other words, the business cannot achieve the desired goals without removing or moving the constraints. Applying automation in certain areas may have a more beneficial effect than in other areas. Matching the Business Strategy with the Automation Strategy by identifying critical constraints is the essence of a CIM strategy.

ARCHITECTURE

Once the CIM strategy has been established, it is possible to address the issue of implementation and the use of technology. A wide variety of technology is available. However, not all is appropriate if integration is to be achieved. An architecture is a framework for representing technology and for ensuring that the technology can be integrated, even within a changing environment. The framework must be able to handle both material and information within the organisation. A concept of layers is needed in order to provide the necessary flexibility for integration. The CIM Architecture is illustrated by applying it to information processing technology. In this case we identify four basic layers: Networking, Data Management, Application, and the User Interface. Each layer is separated from the next by an interface. A protocol represents the functionality of the layer which is supported by services from the operating system.

The Networking Layer is the most mature layer of the architecture. It provides message and file transfer services in a possibly heterogeneous network of computers. The OSI model represents the generally accepted internal structure of this layer. The ability to integrate networks from different vendors is handled today using gateways, either to an OSI Standard network, such as MAP or directly between proprietary networks.

The Data Management Layer is concerned with the management of data in a distributed environment. This layer is concerned with issues such as data ownership, integrity and concurrency control. Data Management is a complex issue, even in a bounded environment such as NC file management. A central database is an idealistic approach which does not reflect normal operations. Furthermore, data may be physically represented in non-electronic form (e.g. on paper). The data management philosophy will be implemented in different ways across the organisation. This must be done, however, by respecting rules concerning data standards, security and authorization. These represent the protocols of the Data Management Layer.

The Application Layer allows different applications to coexist and exchange data. The standards for this layer have to do with the meaning of data, which is determined by how it is used within the organization. It is unlikely that more than basic industry standards can be achieved in the short term. These will take the form of data definitions that can be commonly used by a number of applications with overlapping functionality. IGES is an example of such a standard. Although of limited functionality it does allow the exchange of data between different design packages.

The user interface layer provides the unified user view of the CIM system. The user interface, itself, divides naturally into four components. One is the user's model, the conceptual model formed by the user in dealing with the information manipulated. The second component is the command with which the user interacts. The system commands vary with the user's needs. The third component is the feedback, with which the user is assisted to run the application. Feedback helps the user to be sure that commands are accurately received and understood. The fourth component shows the state of the manipulated data (presentation). The user interface is the least mature layer from the point of view of standards. This is because technology is changing rapidly, and is making the user interface more user oriented than it has been in the past.

ARTIFICIAL INTELLIGENCE

Artificial Intelligence covers a wide range of techniques and technologies at the leading edge of computer science. It is natural to expect that it will have a significant effect on CIM. The focus of this section will be on Knowledge Based Systems which will be positioned in terms of the CIM framework.

The CIM methodology addresses the question of what is CIM. Its methodology can be viewed as a way of integrating a need with a means to satisfy the need. In the above example the need was expressed in terms of a business model and the means in terms of the enterprise model. This is a type of translation problem which can be seen in other areas, too (e.g. translation of a users requirements into a computer program). On the other hand, the CIM Architecture describes how CIM is implemented.

The CIM Architecture views the integration problem as a number of layers linking the human world to the world of machines. This is another type of translation problem (e.g. linking users' commands to the execution of a program).

Artificial Intelligence is concerned with these fundamental questions.

The methodology discussed above has three basic phases (Modelling, Description, Control). Expert Systems can be helpful in selecting models to fit a particular situation. An example of this is the use of XCON to validate configurations of VAX and PDP computers. It is based on a general model of these types of computers.

The ESPRIT project COSIMA is to investigate the use of Expert Systems as application generators for production control systems. This work is based on a generic model of production control which can be applied to different manufacturing environments.

Complex problems often defy the creation of a model, and one has to resort to classification without considering cause and effect relationships. This is the surface reasoning technique which has been applied in many successful Expert Systems, in particular to structural analysis, (SACON) and computer networks, (DART). The application of expert systems to the operational decisions made in manufacturing is covered by the broad definition of decision support systems. These systems aid decision making by exploring alternatives. They have been applied to customer order scheduling (ISA), and capacity planning and inventory management (IMACS).

A Knowledge Based System has an architecture which fits neatly into the framework discussed above. It has three major rule driven facilities. These correspond to the user interface, application, and data management layers. These are the language handler, the interface engine and the database management system. Conceptual schema are shared by the language handler and inference engine. The user interface and application layers are, therefore, the most profoundly affected by AI within the CIM architecture. The use of natural language interfaces and the total separation of algorithms from the control flow, are characteristics of knowledge based systems. AI technology will have a major impact on the definition of the CIM Architecture.

ARTIFICIAL INTELLIGENCE IN MANUFACTURING
T. Bernold (editor)
Elsevier Science Publishers B.V. (North-Holland)

EXPERT SYSTEMS IN MANUFACTURING

FRITZ BURKHARDT
ENGINEERING AND DEVELOPMENT
Ex-Cell-O GmbH, Eislingen, F.R.G.

On a Flexible Manufacturing System presently under construction the procedure for planning such a system is to be described and the functions to operate the system will be explained for the purpose of showing realized elements of expert systems within the control as well as pointing out further needs for expert systems to aid the efficient operation of Flexible Manufacturing Systems.

INTRODUCTION

Over a period of 2 years a Flexible Manufacturing System (FMS) for the production of housing parts for hydraulic pumps was developed, first under competitive conditions, and later in a specific engineering study.

A team of experts conducted thorough research of possible variants, worked out influencing factors and interactions, formed an opinion relative to advantages and disadvantages and the degree of fulfilment of the given requirements. This brought about a system that to our knowledge is presently unique in size, depth of implementation and complexity.

The system is in component assembly. Beginning of 1986 a cell with all major components will be in operation for testing the computer software.

The Manufacturing System is designed to produce 1 Mio parts per year with 14 presently known variants with varying manufacturing content. The production is planned to run around the clock during 250 days per year in partially unmanned operation.

The goal is to reduce cost by 24 hour usage, built-in flexibility and short change over times, minimum number of parts in the system and "just in time" delivery to the customers.

1. Planning of the FMS

1.1 Definition of Goals

The planning of such equipment requires a careful definition of goals, requirements and all influencing factors such as:

- Establishing types of parts and production quantities
- Capacity requirements
- Working hours per year
- Establishment of technical requirements
- Establishment of requirements for the computer control
- Definition of the priorities and recognition of the interactions between goals and priorities:
  - . Flexibility
  - . Interchangeability
  - . Redundance
  - . Manless operation
  - . High availability
  - . Minimum scrap rate
  - . Capacity oriented production
  - . Assembly oriented production
  - . Multiproduct manufacturing
  - . Minimum change over cost
  - . Minimum number of parts in the system
  - . Capacity compensation between running orders

1.2 Conception

1.2.1 Layout and Optimization of the System

The goals and given priorities suggest the use of a Flexible Manufacturing System.

The high production, difficult to combine with flexibility and interchangeability, demands particular consideration in the selection of the equipment for manufacturing and part transport.

Picture 1 shows the configuration of the system with the major components.

The transport system with rail-guided directly addressable vehicles and maximum transport speeds and the centrally located work piece loading and clamping stations determine the arrangement of the layout in line.
The working stations to be serviced are arranged on both sides of the central transport in such a way that from the opposite side optimal access for service and tool loading is provided.

1.2.2 Working Stations

After determination of the manufacturing process the heart of the system, namely the working stations, wants to be attended.

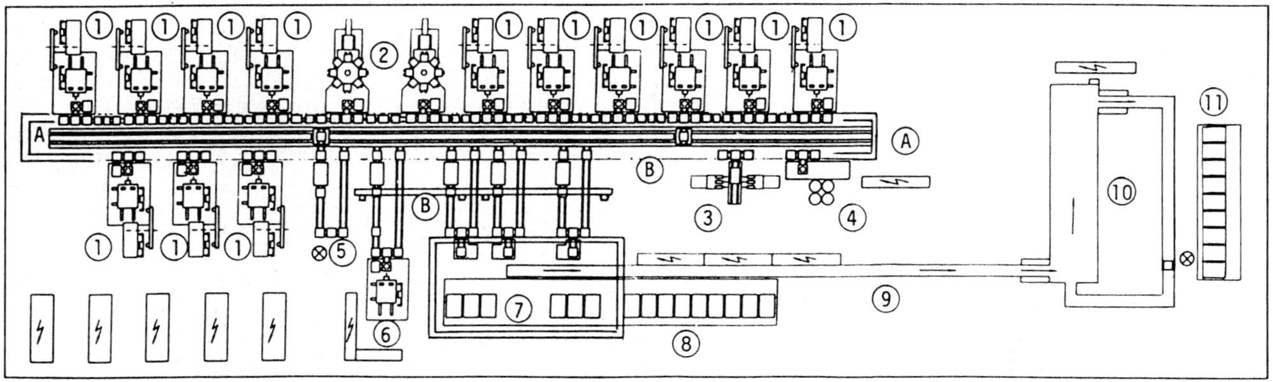

1 UNIVERSAL MACHINE UNITS
2 MULTISPINDLE UNITS
3 FINISH BORING MACHINE
4 BUSHING PRESS
5 MANUAL LOAD STATION
6 GAGING AND LASER MARKING
7 AUTOMATIC LOAD & UNLOAD
8 CASTINGS DISPOSITION
9 FINISH PART TRANSPORT
10 WASHING MACHINE
11 FINISH PART STOCK

A RAILTYPE TRANSPORT
B GANTRY TYPE TRANSPORT

Ex-Cell-O GmbH

PICTURE 1: FLEXIBLE MANUFACTURING SYSTEM

For interchangeability it is desirable to have as many equal universal machines as possible to complete all necessary operations. The flexibility of the universal units permits a favourable distribution of the orders running simultaneously over the available manufacturing capacity. In the event of failure the task can be taken over by other units.

Technical reasons also require dedicated units with little or no interchangeability such as multispindle units, bushing presses, finish boring machines as well as gaging and laser marking units.

This completion in various steps multiplies the number of transport orders because now there are also transports between machines and buffer stations.

1.2.3 Transport Systems

The total transport system consists of several complementing or replacing sub systems.

A high performance central transport system facilitates the transport of the fixture pallets.

By using one gantry-type and two rail-type vehicles the transport can take place in different levels and, therefore, serve complementary as well as overlapping.

Further systems are for the supply of castings from delivery to clamping station, and finish part transport from clamping station via the washing machine to the manual unloading station.

There are additional conveyors between clamping station and central transport as well as another gantry for integration of the gaging machine.

The transport control will be explained later in more detail.

1.2.4 Material Handling and Material Flow

To assure safe and simple part handling, the various types of parts need to receive same clamping and gripping points.

The parts are delivered, sorted and oriented on Euro-pallets.

Possible wrong parts are sorted out by a vision system prior to loading into the clamping fixtures.

Loading and unloading of the fixtures is taking place with two gantry-type robots.

For quick tracing of faults each part is laser marked.

After completion of manufacturing and gaging it will be marked with identification number as well as date and number of the work stations involved in manufacturing.

1.2.5 Simulation

After establishing the manufacturing process, and after determining the machining stations and the transport equipment, validation and optimization of the system with computer-aided simulation is necessary to follow.

The complexity of the dynamic proceedings in a Flexible Manufacturing System affords the use of simulation as an important aid.

For this purpose we have implemented a software package on our CAD computer that permits us to obtain an optimal layout in 3 steps.

Step 1

Through static calculations the feasibility of the system under given conditions is verified.

The results of step 1 for example are:
- the number of machines required and their statical loading
- the approximate number of fixture pallets
- the number of transport systems and their statical loading

These results can be optimized by an iterative process.

Step 2

Based on the data obtained in step 1 the dynamic behaviour of the chosen arrangement of machines, pallets and transports is simulated with best possible approximation of the proceedings for a particular mix of parts and orders.

The following results are obtained:
- dynamic loading of machines and buffers
- dynamic loading of the transport
- total production obtainable
- work-in-process time
- number of pallets per part type

The material flow in the FMS depends to a great extent on the number of fixture pallets in circulation.

In a further step the optimum number of pallets can be determined using the results obtained through evaluation of step 2.

Step 3

With a third software module the flow of pallets in the system can be visualized by animation.

Bottle necks or idle phases show critical points for system optimization.

With this knowledge we can go again into the iterative process of step 2.

A good basis for a computer controlled FMS is created as soon as the system has been optimized according to its requirements, and a good relation between number of pallets and work-in-process time has been obtained.

1.2.6 Tool Management within the FMS

For each cutting tool within the system an identification number is assigned after the presetting procedure. Additional information like offsets, tool life data and number of duplicate tools in storage are associated with the identification number and stored in the computer.

For storage each tool is carrying the identification number on a bar code label or, more recent, on programmable chips.

When the tool coming from stock is loaded into the machine storage, the identification code is passed into the machine control and the associated data called for from the computer.

Data of preset tools, both in stock and in the machine storage, are managed by the computer. On request, tooling lists are issued by the computer for orders following.

1.2.7 Quality Assurance

The strategy is aimed primarily at quality assurance by a proper presetting procedure and monitoring the tools during the cut.

The following measures are taken:

- Use of tools with high life expectancy
- Safe presetting within the tolerance range
- Tool monitoring for dulling and breakage in the machine
- Tool life monitoring

Dimensions with very close tolerances are post process gaged on a gaging machine within the system and the results statistically collected.

On a univeral inspection machine complete checks are carried out at random.

More details on this subject are referred to in chapter 2.1.2.

1.2.8 Auxiliary Equipment

For the operation of a Manufacturing System further equipment is required:

- Coolant equipment
- Chip removal system
- Part cleaning and washing
- Safety equipment.

1.2.9. Process control

The host controller with the process control software is the brains of the FMS and must satisfy very high requirements.

The large number of system components in connection with short machine cycles and resulting high transport frequency demands very powerful hardware which can satisfy a very high number of events with shortest reaction times.

In the hardware configuration as per picture 2 all hardware modules required are shown as they are foreseen in the various areas of operation.

The hierarchical setup and the connections between the individual modules are indicated.

There are various levels with subdivided responsibility and competence. On the host level it was considered important to implement redundancy.

The software modules employed in the FMS and the functions realized are shown in picture 3.

2. ELEMENTS OF EXPERT SYSTEMS

The foregoing introduction of the elements of the system served for better understanding of the equipment and the tasks of the system host computer.

In the following, several software modules will be explained in more detail to point out possible elements of expert systems as well as the potential usage.

2.1. Realized System Elements

2.1.1. Transport control for fixture pallets

The whole material flow control consists of several sub-systems with separate decision criteria:

- Casting supply station
- Clamping stations with coming and going conveyors
- Central transport to working stations and buffers
- Gantry transport between gaging station and unclamping stations
- Finish part conveyor

The diagram in picture 4 shows the flow of information and material between the sub-systems.

In the following we wish to eyplain the transport control for pallets in more detail.

In picture 5 the decision criteria for transport control are shown.

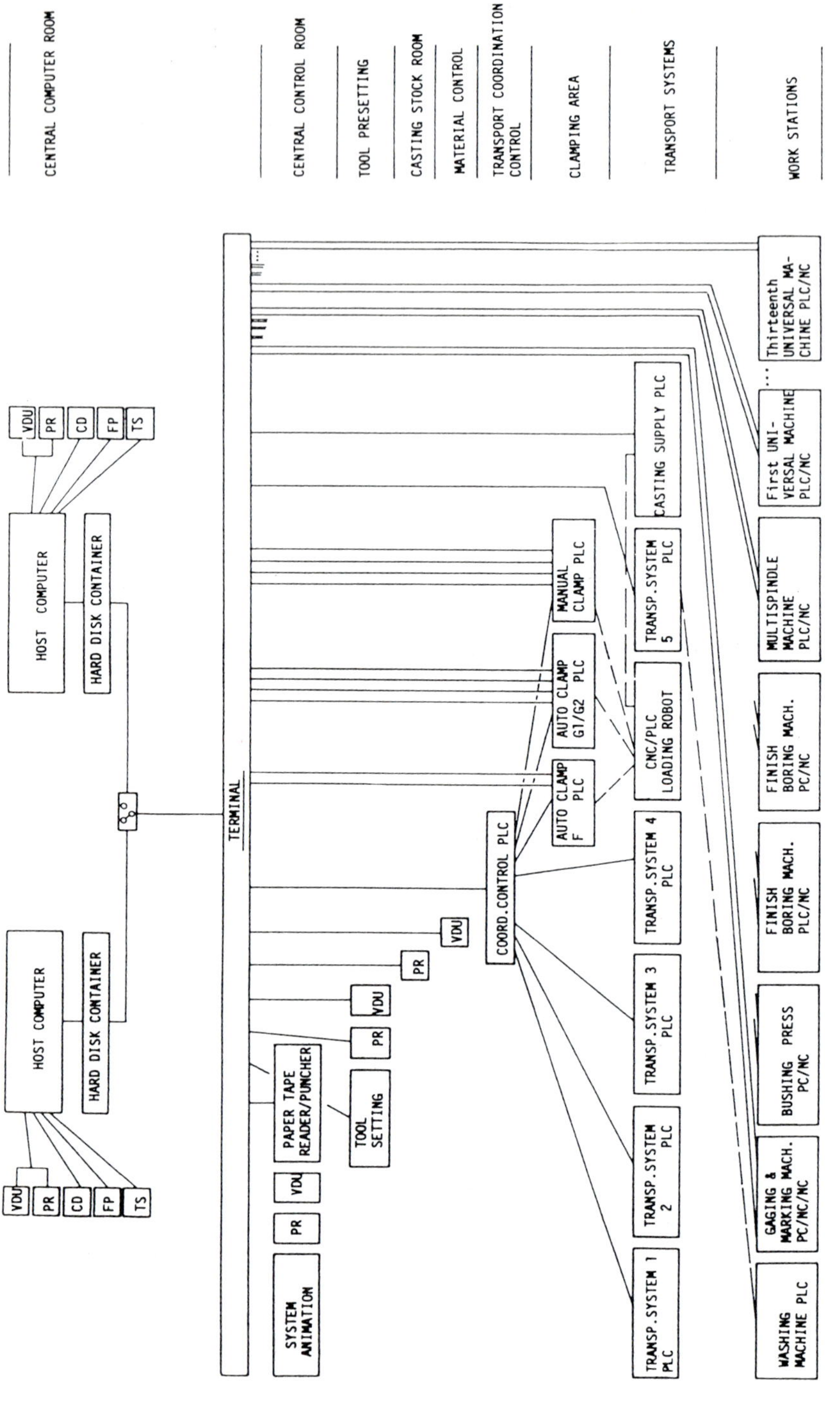

VDU = VISUAL DISPLAY UNIT PR = PRINTER CD = CARTRIDGE DISK ——— = SERIAL INTERFACE
FP = FLOPPY-DISK TS = TELESERVICE - - - - = I/O INTERFACE

PICTURE 2: HARDWARE - CONFIGURATION

Ex-Cell-O GmbH

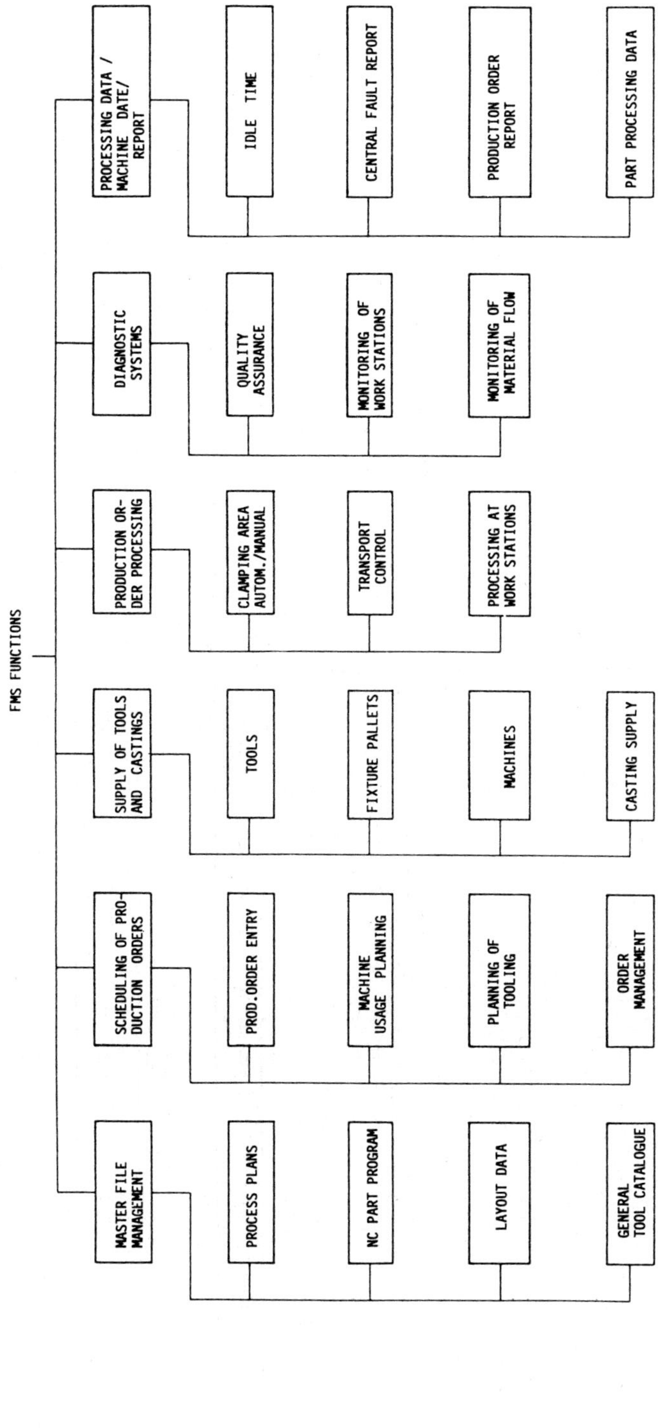

PICTURE 3: FMS FUNCTIONS

Ex-Cell-O GmbH

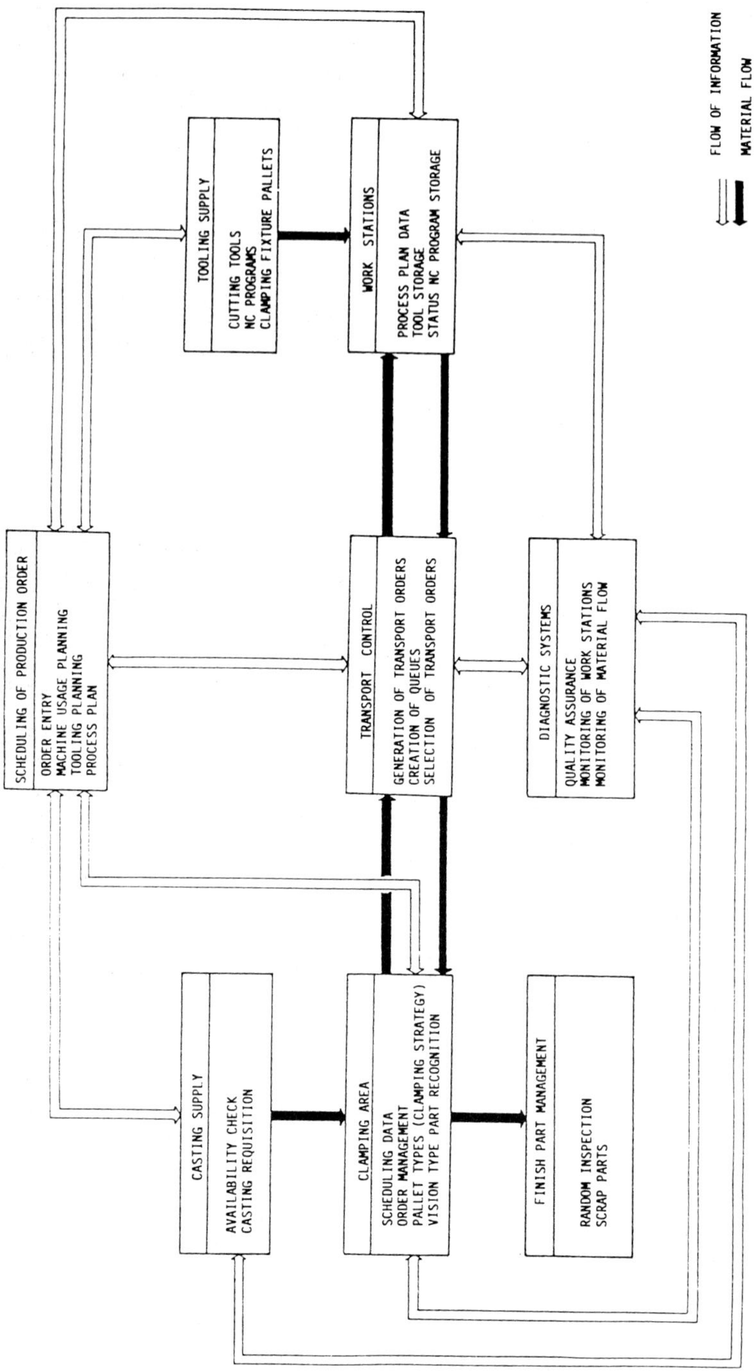

PICTURE 4: PROCESS CONTROL

Ex-Cell-O GmbH

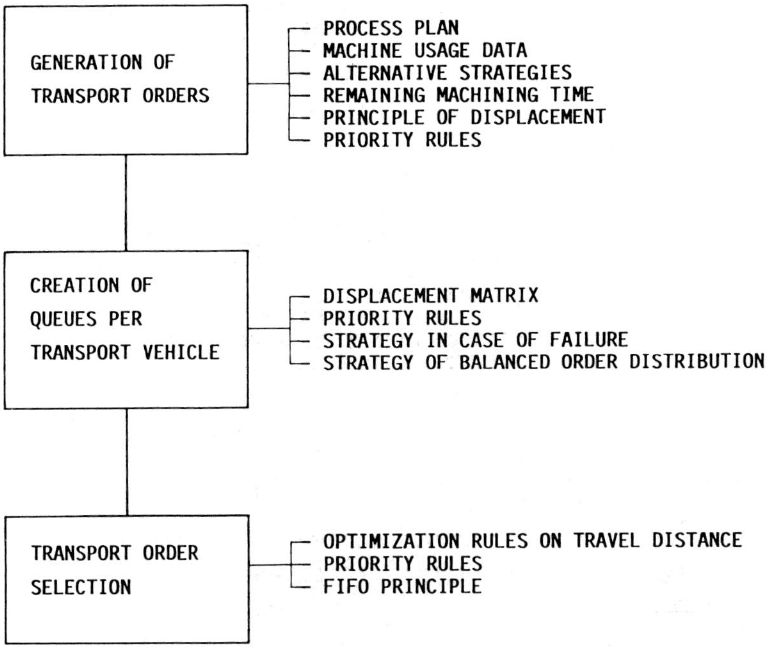

PICTURE 5: DECISION CRITERIA FOR TRANSPORT CONTROL

Ex-Cell-O GmbH

Generation of Transport Orders

The transport order is generated from the signal to load or unload a work station.

Basis for establishing a transport order is the process plan and the data from the machine usage plan.

The process plan describes the association of the individual machining processes to the work stations in the proper sequence.

We differentiate between preferable machines and alternative machines. In case of machine failure alternative routes can be selected depending on the present system status.

If per machine allocation plan several interchangeable machines are free for loading, the remaining machining time will be the decision criterium.

If a finish machined fixture pallet is ready for unloading from a work station and no successive station is free to accept, the transport is delayed until a defined limit is reached. After this the pallet is moved into the buffer. This principle of displacement is used to relief the transport system by reducing moves to the buffer.

For changes in production rates deviating from the given part mix, priority rules are employed.

Creation of queues per transport vehicle

The assignment to the particular transport system is in accordance with the source-target definitions.

The required information is laid down in a displacement matrix.

For transport orders that can be carried out by several vehicles, priority rules are used.

Signing in of orders into the vehicle specific queues is dependent on the availability of the vehicles.

If in case of failure alternative vehicles are available, the transport order can be moved to another vehicle.

Order assignment also takes into consideration if a vehicle is overloaded due to high number of transport orders. In such a case it is tried to balance the order backlog within the queues.

Transport order selection within the queues

To relief the transport vehicle there are various optimization rules on travel distance. For instance, after unloading the fixture pallet at a certain work station the vehicle is, again, loaded with a finish part pallet.

It also appears to be useful to provide certain stops with priority to obtain a complete transport cycle. If the cycle needs to be interrupted the desired pallet is preferably picked up from the buffer.

In case none of the above mentioned criteria can be fulfilled the next transport order is selected by using the FIFO principle.

The description of the transport system serves to show the structure of the main decision criteria to be considered between a transport signal and a transport order.

It also gives an impression of the complexity of the order strategy, emphasizing the necessity of careful process planning in order to comply with all system requirements in the best possible way.

2.1.2. Diagnostic Systems

We distinguish between process control and process supervision. Both can be considered regulating circuits continuously trying to achieve a state of balance.

Each individual signal, be it a transport order or a fault signal, calls for reaction in the control system. Every change of a given situation initiates a diagnosis to identify the optimal action to be taken by going through the branches of the decision criteria. The results are for example: Transport orders, tool exchange or failure strategies.

Process supervision consists of various diagnostic systems:

- monitoring of work stations
- monitoring of material flow control
- monitoring of part quality.

In the following we wish to describe the system of quality assurance in more detail.

Quality assurance is subdivided into monitoring of tools to prevent scrap, and product monitoring to recognize scrap.

Picture 6 shows the criteria for decisions.

2.1.2.1 Monitoring of tools

- Tool presetting

  Actual dimensions are measured on a tool presetter which is coupled to the computer. They are compared with the tool reference data in the master file. Depending on tool type they are either adjusted or the offsets are stored in the memory.

  After gaging, the tool is assigned with an identity number, coded and then transported to the tool storage or to the machine storage.

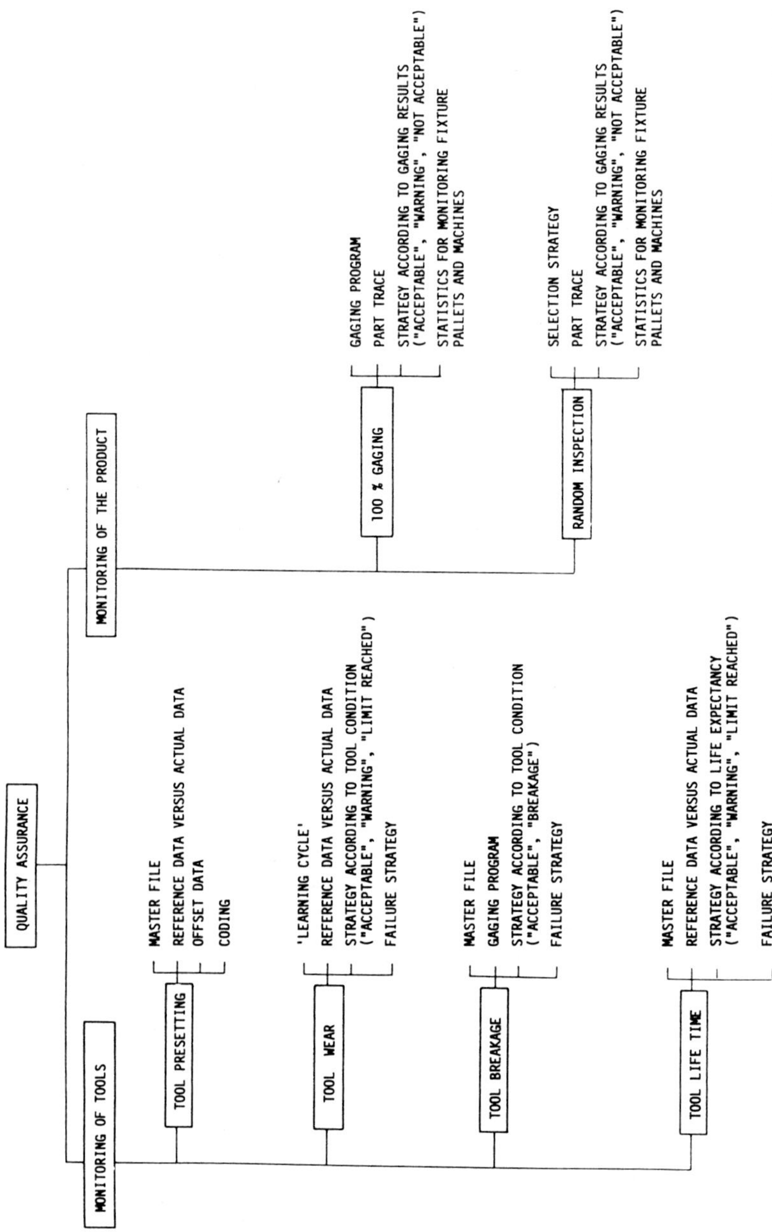

Ex-Cell-O GmbH

PICTURE 6: DECISION CRITERIA FOR QUALITY ASSURANCE

When loading the tool into the machine the offset data and tool life data stored in the computer are transferred and per this coding assigned to the respective tool.

- Monitoring of tool wear and tool breakage

  Each machine tool is equipped with a System A to monitor wear and breakage during the machining process. System A is limited to certain tool types and tool sizes.

  The relevant tools are taken through a "learning cycle" in order to establish the basic values and limits for tool wear and tool breakage.

  The comparison between reference data and actual data occurs in each cycle during the cutting process.

  The strategy according to the condition distinguishes between "acceptable", "warning", and "limit reached".

  When in "acceptable" the machining process continues; in case of "warning" a signal is given and a protocol indicating type and location of action to be taken is shown.

  If the "warning" limit is reached or exceeded, or in case of breakage the machine is stopped. If available, a duplicate tool is used.

  To complement System A a second System B is provided for small tools to monitor tool breakage after machining,therefore, requiring additional time. Furthermore, there is no "warning" limit.

- Monitoring of tool life

  Until reliable monitoring of tools during cycle time is obtained, a monitoring program for tool life is carried out in addition.

  The minimum tool life value arrived at empirically is stored in the master file and transferred to the machine.

  At each use of the tool the time of engagement is registered and the value deducted from the reference data. Analysis and strategy are similar to the above and are processed accordingly.

2.1.2.2 Monitoring of the product

- 100% gaging

  Here certain critical dimensions with close tolerances are gaged on all parts.

  The gaging program is layed out according to the criteria in the inspection plan:

  - gaging reference data and tolerance limits
  - tool type
  - number of operations.

  The part is traced according to the specific machining sequence.

In the strategy according to gaging results we distinguish between "acceptable" and "warning" like in monitoring of the tools. In case of scrap the machine stops, the tool which caused the failure is traced, and the respective work part is marked and ejected.

Statistics provide us with the respective conclusions in regard to quality of tools, machines and fixture pallets.

From the gaging data and the sequence of operations stored in the computer a part trace is obtained admitting a quick reference to the source of failure as well as to trends.

- Random inspection

By means of random selection certain parts are gaged completely on a universal gaging machine in order to obtain statistical data for those dimensions which are not checked in the 100% gaging test. These are dimensions where sudden and random deviations are not expected.

The strategy of selection foresees regular inspection of the results of all machines.

Gaging evaluation follows similar criteria as in "acceptable", "warning" and "not acceptable".

With the help of statistics on frequency of events, respective conclusions are drawn in regard to the machines.

The chapter 'quality assurance' shows the effort exerted for monitoring manufacturing in automatic production.

It is supposed to give an impression on the numerous decision processes going on to determine whether we are producing good parts or if we are in the "warning" range requiring quick action, or if the "red" light goes on indicating that the machine has stopped and certain failure strategies have to be initiated.

2.2. Further needs for expert systems

While in the previous chapters executed controls were explained which possibly contained elements of expert systems, we now would like to point out another possibility for application where we are sure users of FMS would appreciate support, namely for machine scheduling.

Presently known systems accomplish the machine scheduling by dialogue between the system operator and the computer.

The steps of the iterative process are depending on the complexity of the task.

Even simple systems can be confronted with complex scheduling which do not permit an optimal machine loading without additional support like for instance dynamic simulation.

In a large system just an order revision can necessitate an extensive rescheduling run.

Difficulties arise particularly when a system is utilized in a full three shift day without any spare capacity, and any disturbance will affect the delivery date.

As pointed out in chapter 'simulation' there are many factors of influence and interactions between transport, work stations, number of pallets, buffers, sets of tooling available, work load, finish dates and running orders.

Because of the complexity pointed out we are possibly approaching the limit of human capability.

This has negative effects on the utilization and, therefore, the efficiency of a Manufacturing System.

The scheduling should be an automatic process which will be started by either a new order, an order revision or a machine failure, that is any change to the system capacity.

In case of machine or component failure the optimal failure strategy should also be selected automatically.

ARTIFICIAL INTELLIGENCE IN MANUFACTURING
T. Bernold (editor)
Elsevier Science Publishers B.V. (North-Holland)

GUIDING TECHNIQUES FOR MANUFACTURING SYSTEMS

Guy DOUMEINGTS
Deputy Director
GRAI Laboratory
Université de Bordeaux 1
33405 Talence Cedex - France

This paper deals with the use of computers in factory management. We present a proposal for designing computer structure in order to guide (or to control) the manufacturing system towards its goal. In the first section, we will give the outline of GRAI method which enables to design manufacturing system, acquiring the knowledge and the expertise from the users. In the second section we will define the guiding concept we use in our approach and how we use a standard approach to solve the problem. In the third section we will show how we have introduced Artificial Intelligence Techniques into our approach.

INTRODUCTION

The use of computers to control manufacturing systems is certainly one of the most complex task in the field of Computer Integrated Manufacturing.

Our laboratory has been carrying out some research in this field for over twelve years. It appeared very quickly that :

- it was necessary to use a method in order to analyze the manufacturing systems and elaborate their specifications.

- the users were not satisfied with the running of several functions particularly in scheduling, planning and feedback from the shopfloor to the various levels of factory management.

Besides we have noted the dynamic aspects of manufacturing : the manufacturing systems keep evolving over the time and many events occur during the running of the systems altering thus the state of these systems.

Therefore we have searched for methods and techniques which would allow us to design manufacturing systems in an open way. The use of Artificial Intelligence Techniques looked quickly as a potential solution.

We distinguish three stages when using Artificial Intelligence Techniques :

- First, knowledge represention : how shall we represent the manufacturing systems, the structures and the various elements ?

- Second, problem solving : How shall we reach the objectives given to the systems ?

- Finally, the third stage is the development of the application : It may be performed either with an A.I. language like LISP or PROLOG or a knowledge base language like KEE, ART, KNOWLEDGE CRAFT ...

The GRAI Laboratory's research has been carried out according to these three stages and our presentation will resume this development.

We will first describe manufacturing problems and the GRAI method based on the GRAI conceptual model.

We will then show how we solve the problem of manufacturing control. We are developing a computer tool called "guiding module" which allows us to implement in the the computer the model we have designed.

We have lately decided to introduce Artificial Intelligence languages into our development in order to improve the flexibility of the model : this issue will be the third section of our presentation.

1 - MANUFACTURING PROBLEMS AND GRAI APPROACH : A KNOWLEDGE REPRESENTATION

In the field of manufacturing the works of our laboratory deal with the design and the control of Production, Factory Management (Fig. 1) and more recently Flexible Manufacturing Systems (FMS).

The main conclusion we can make about the computerization of the management of such systems is that it is difficult to reach a satifying stage from the users' point of view. We think that this is mostly due to the fact that we donnot obtain a realistic model.

Therefore the GRAI Laboratory has developed a method to design Production Control Systems (PCS) and FMS. This method is called GRAI method (Graphs with Results and Activities Interrelated).

It is based on a conceptual model, it uses a graphic tool and a structured approach to collect the information and to elaborate the models.

The GRAI conceptuel model [DOU 84.11], [DOU 85.3] is represented in Figure 2. It shows with abstract terms the structure of the manufacturing systems, the various entities and the links between them. The GRAI Conceptual Model has been derived from several theories : the organizational theory [SIM 60], [SIM 69], [SIM 75], [LEM 77], the hierarchical system theory [MES], the discrete activity theory [PUN 84].

The Conceptual Model of Manufacturing System is split up into three subsystems : Physical System (PS), Informational System (IS), Decisional System (DS).

The Decisional System is divided into a hierarchy according to two criteria :

- a criterion of time : horizon of decision making and period for revising these decisions ;

- a criterion of function : determined according to basic functions of production : to design, to purchase, to plan, to manage the facilities (or to manufacture), to control and to deliver.

We obtain a division into Decision centers (Fig. 2). For the description of each DC, we have chosen the concepts of activity (with different types of activity : physical, informational or decisional).

The graphic tools are mainly the GRID which shows the structure of the decisional system (Fig. 3) and the GRAI nets (Fig. 4) which enable us to describe each decision center [DOU 84.11].

The structured approach called the GRAI method is divided into two phases : the Analysis phase and the Design phase (Fig. 5). Several actors must be involved in this approach :

- The synthesis group : It leads the project and validates the results. It is composed of the main executives of the manufacturing system.

- the analysis group : It collects the information from the manufacturing systems and is a true sensor embedded into the system.

- the specialist group : specialists in the GRAI method, it helps the analysts in gathering the collected information under GRAI form.

- the group of interviewed people : this group is composed of executives who have a decisional function at any level of the structure.

In the analysis phase we collect all information from the system. We thus gather the expert knowledge of the factory management people.

In the top down analysis we determine, with the synthesis group, the content of the Grid (Fig. 6) : Horizon/Period, Decision Centers (DC), links between DC (The relationship between two DCs is represented by a double arrow and the main exchanges of information by a single arrow).

In the bottom up analysis, we start with the lower levels of the grid and analyze through the GRAI nets the detailed activities of each DC. We thus collect :

- the data necessary for the decision making,

- the data related to decision frames :

. decision variables,
. decision operators,
. decision rules,
. constraints
. criteria,
. objectives.

This analysis is performed in a static way (when all information and supports are present) and in a dynamic way (in a perturbated state when a support is missing).

In such a study the executives within the manufacturing systems must be involved in the synthesis group. We are aware that these executives donnot have a lot of time that is why the structured approach and the method have been specially designed in order to take only two or three hours of their time every two weeks during a period of four to six months. The result of this analysis phase is detailed in an analysis book in which we collect :

- one or several grids, according to the complexity of the systems. We usually have an aggregated grid and several detailed Grids (Fig. 6).

- several GRAI nets : each square in the Grid has a GRAI net which shows the detailed decision activities of this Decision Center.

During the analysis phase and thanks to the conceptual model and the formal rules we have set, we are in a position to detect any inconsistencies. These inconsistencies deal with :

- information,
- structure of decision centers,
- coordination between decision centers,
- dynamic behavior.

Examples of inconsistencies related to :

- the information systems :

lack of updating of the information,

. lack of previsional information,
. delay in the transfer of information,
. redundancies of information,
. Inconsistencies between units of measure,

- the structure of the decision centers :

 . consistency between the period of information updating and the period of decision making,
 . lack of decision frame,
 . lack of performance evaluation.

- the coordination between decision centers :

 . when one Decision Center must comply with two different Decision Frames,
 . when the decision frames skip several levels,
 . inconsistencies between the horizon/period of each level.

- the dynamic behavior :

 . analysis of decisions in an altered situation,
 . search for missing elements.

Once the inconsistencies have been detected and the objectives given by the top management (strategic objectives, previsional development of manufacturing systems), we start the design phase.

The approach follows also a topdown and bottom up design :

- During the topdown design we design a new grid in order to solve the inconsistencies, we use the design rules given by GRAI CONCEPTUAL MODEL ;

- During the bottom up design we follow two steps to design GRAI nets :

 . First the MACRO-GRAI gives us the overall relationship between decision centers,
 . Second, the detailed GRAI gives us the final design.

Once the design phase has been carried out, we are in a position to write the specification book.

So GRAI CONCEPTUAL MODEL may be considered as the foundation of knowledge representation and GRAI method as the technique to acquire the knowledge from experts.

The work carried out at present at the Laboratory and in several companies which use

GRAI method (particularly VALORIS, from SGN group) leads to the development of a computer support for the method and of expert systems for the diagnosis of the grid.

2 - **MANUFACTURING CONTROL BY "GUIDING TECHNIQUES" : A PROBLEM SOLVING SOLUTION TYPE**. [BER 85.8], [DOU 84.11]

This section deals with the control of manufacturing system particularly at workshop level (we are located at the bottom level of the Grid and we are dealing with the column TO PLAN).
In the previous section we have described the method for designing the model of manufacturing unit, mainly the control of this unit. This model is described with GRAI tool.

Before presenting manufacturing control by guiding techniques, we must define the "guiding concepts" and then determine a structure for a computer guiding system.

2.1 - Guiding concept for manufacturing systems

Guiding a general system consists in monitoring its running and driving each part of the system towards its objectives.

Manufacturing systems belong to complex systems. In order to achieve its objectives this complex sytem needs to be split up. For each subsystem we must define a sub-objective connected with the main goal of the system.

As above mentioned, guiding a system means driving it towards its goal. So in order to do such a thing, four activities must be defined (Fig. 8) :

- To calculate a reference : this activity determines pathes allowing the system to reach its goal. For a system described as a structure of subsystems, the reference is defined as a set of sub-references. One for each subsystem.

- To control the elements (or subsystems) : this activity is related to the sending orders to elements in order to guide the system towards its goal. It is done according to the reference.

- To follow the element state : this activity as to measure a set of significant variables periodically.

- To react (or feed-back) : this activity compares the true state of the system with the reference. It may decide :

. to adjust orders for system elements to assess the gap between this state and the path defined by the reference.

. to ask for a new reference calculus better adapted to the new state of the system.

2.2 - <u>Computer guiding system</u> :

The basic principle of GRAI problem solving in this case is to implement the model of the manufacturing unit control in the same structure as GRAI model : We use the guiding module (Fig. 9, 10).

A module has an input list and an ouput list. It is related to an activity. In the input list there are tasks to be performed by the activity. For each task, five elements may be noted in the input list : name of the product, date when the product is ready, routing operation number, date when the operation is over.

This description of both lists is only an example to show how a guiding module can communicate with the closest upper level. The running of the guiding module depends on what is set in its input list. And the content of the output list may give the state of the production at each time.

In Figure 11, we can see an example of the use of guiding modules in a simulation. There are three main hierarchical levels. The hierarchy which exists in the real production system will exist in the structured programming. For each type of activity a type of guiding module is defined.

Each level has some predefined functions. These functions are part of the algorithms of the simulation. On the middle level, they control this running. That means that they choose in the input list the product which is to be performed first. Each guiding module is responsible for one activity module. The highest module starts the processing and gives the result to the next lower module. At the last level the simulation algorithm takes place. According to criteria defined by users, it manages the whole set of input and ouput lists. It contains the optimization method chosen during the analysis of the shop.

We represent in Figure 12 a hierarchical structure of a manufacturing system with guiding modules.

For the mentioned application, the developed system has been connected to a GANTT-based color graphic system, each diagram representing manufacturing orders.

From this application, we can derive the following conclusions :

- the model has been perfectly runned by the user who has recognized familiar representations, especially through graphic communication system.

- the response time of the system was very short depending on the introduction of new data, the modification of means or the disturbances taken into account.

- the system did not give the optimal solution but an acceptable solution. The search for new conditions was not automatic but the user contributed its knowledge and acquaintance with the workshop. The results were quite satisfactory..

Even though this application has been successful, we wished to design a semi-intelligent system able to lead the user towards more optimized solutions.

3 - A.I. LANGUAGE AND GUIDING TECHNIQUES FOR MANUFACTURING SYSTEM

The future research progress which our Laboratory will carry out in this field is shown in Figure 13. The goal is to improve the intelligence of our system introducing several modules which interacts with various actors. This intelligent system will allow the scheduler to improve the search for a solution through a pre-analysis and a post-analysis.

The problem we raise deals with the choice for languages in order to develop this system.

The whole guiding structure obeys to artificial Intelligence concepts, but only some parts of it use artificial intelligence techniques.

The first one is the reference calculus module. In it, two functions need artificial intelligence techniques to be solved :

- the choice of a strategy for the search of a scheduling. This choice implies a non-deterministic analysis of the capacity of resources and the manufacturing orders to be performed. The choice is done with a set of rules depending on the manufacturing system. This set of rules evolves along with the manufacturing system (a new product or a change in the machines are examples of possible evolutions). The result of the choice is another set of rules which will be used to search for or adjust the scheduling.

- the search for a scheduling used to calculate the reference planning. It implies backtracking techniques easily implemented with a language like Prolog. The number of possible combinations is reduced by the previous choice of a set of heuristic rules.

Then the feedback module also uses artificial intelligence techniques when it tries to find new solutions from the reference planning and data collecting by the follow-up activity. Its search must also use the strategy previously defined.

All the other modules are designed from deterministic algorithms. For example, the simulation module do not need to be "intelligent". Its only purpose is to assess future states of the guided system from a defined state and according to well-defined rules. This running is entirely deterministic.

Finally the development of object oriented language in order to translate the GRAI conceptual model opens a new field in our research.

CONCLUSIONS

From this approach, the GRAI laboratory has already achieved several guiding software used in industry. Its research is supported by the ESPRIT European Program. Actually, the work is essentially directed towards the use of Artificial Intelligence techniques for scheduling function and the feedback Module.

The choice of languages is not the most important one. More important is the methodology used to design the software. For manufacturing systems, a software design method based on Artificial Intelligence concepts seems able to solve the problem of manufacturing control.

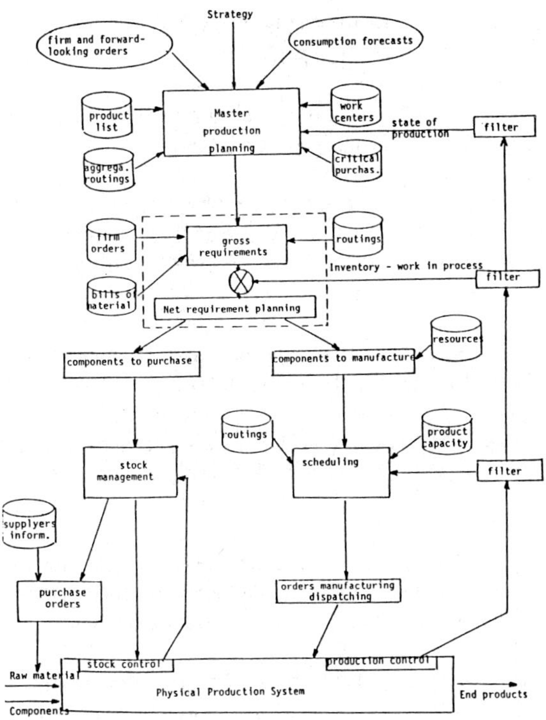

Figure 1

PRODUCTION CONTROL SYSTEM

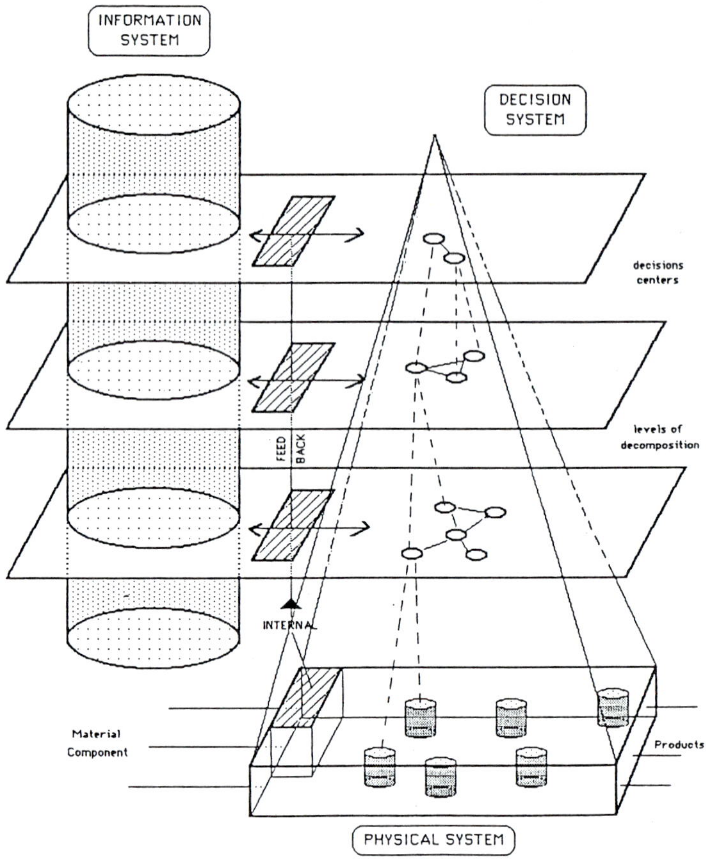

Figure 2

GRAI Model

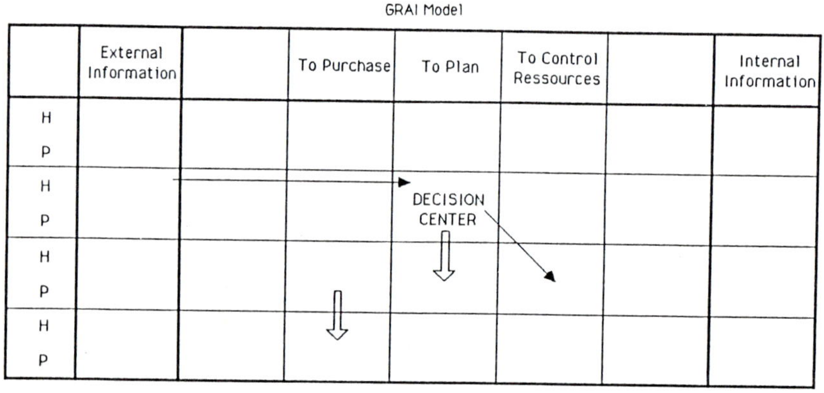

Figure 3

GRAI Grid

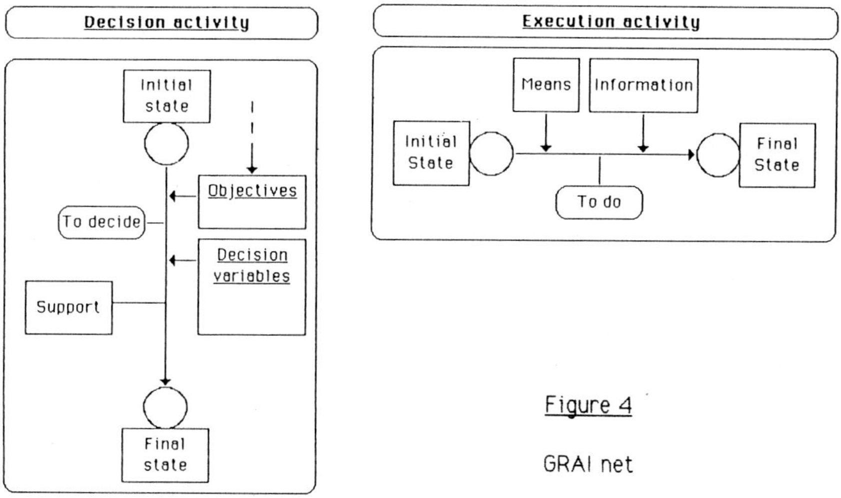

Figure 4

GRAI net

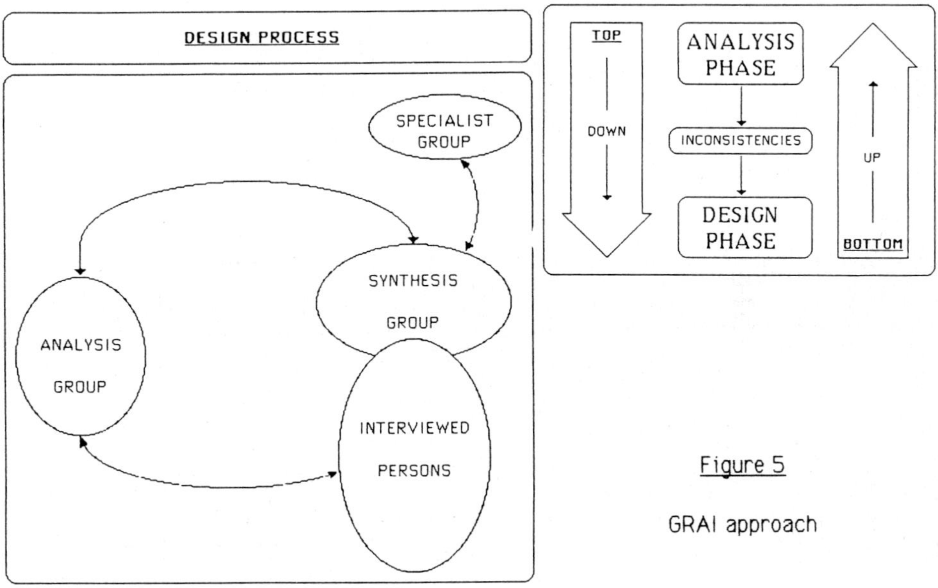

Figure 5

GRAI approach

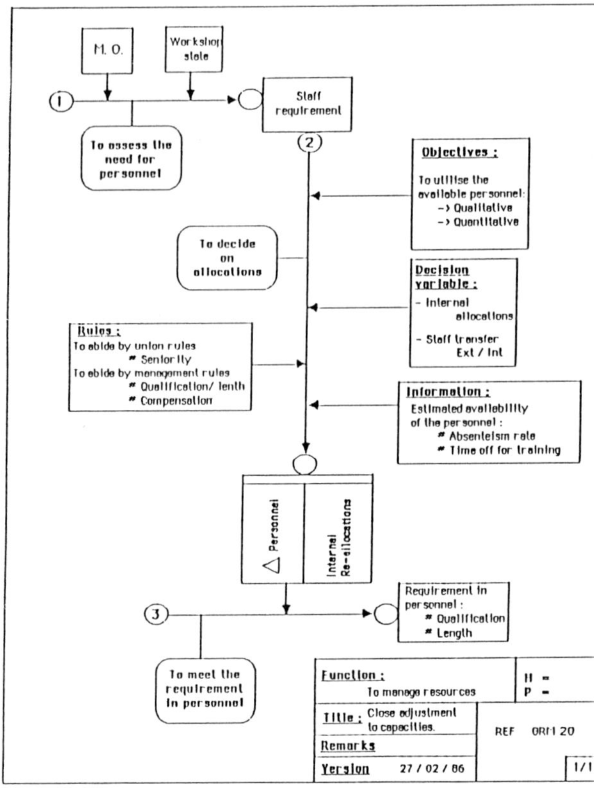

Figure 7

GRAI nets

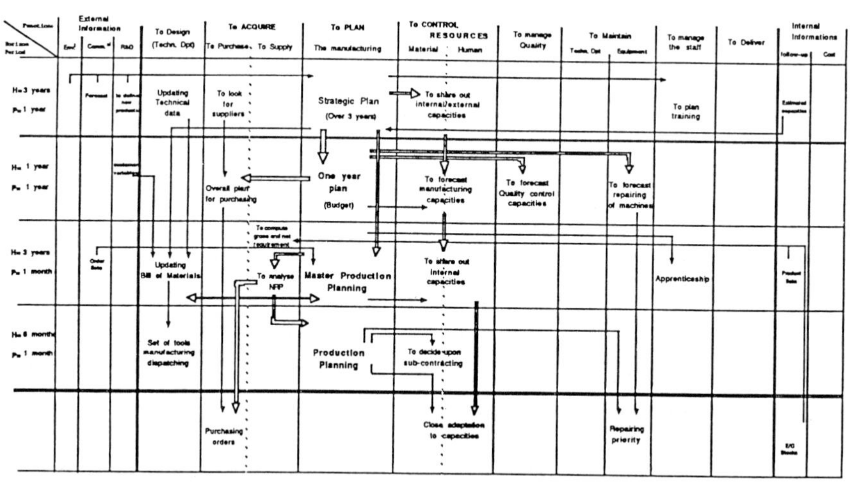

Figure 6 Application : GRAI's grid

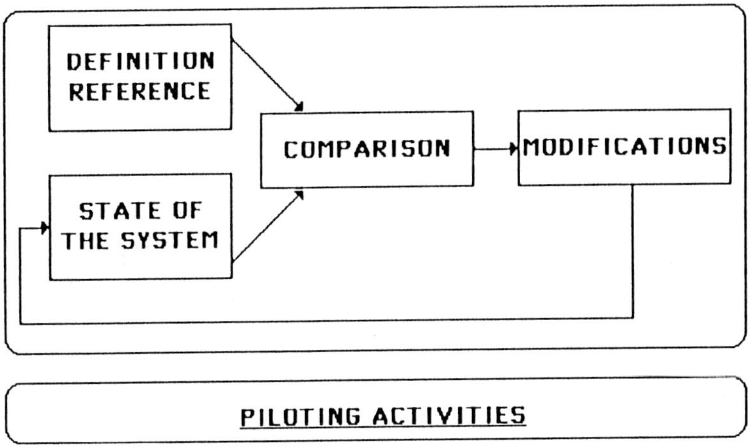

Figure 8

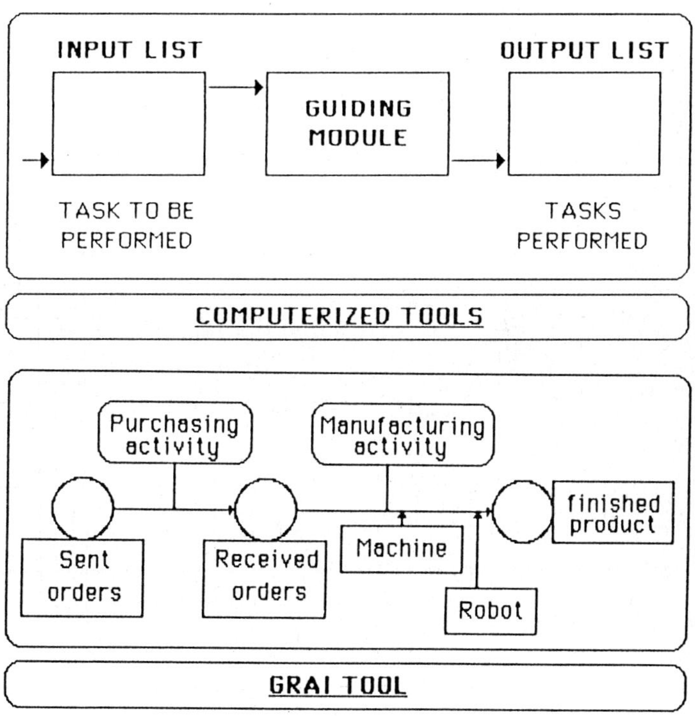

Figure 9

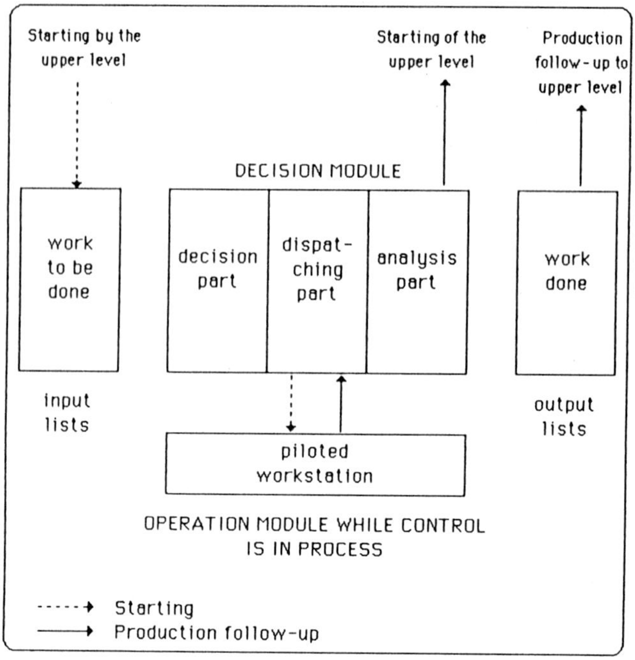

Figure 10

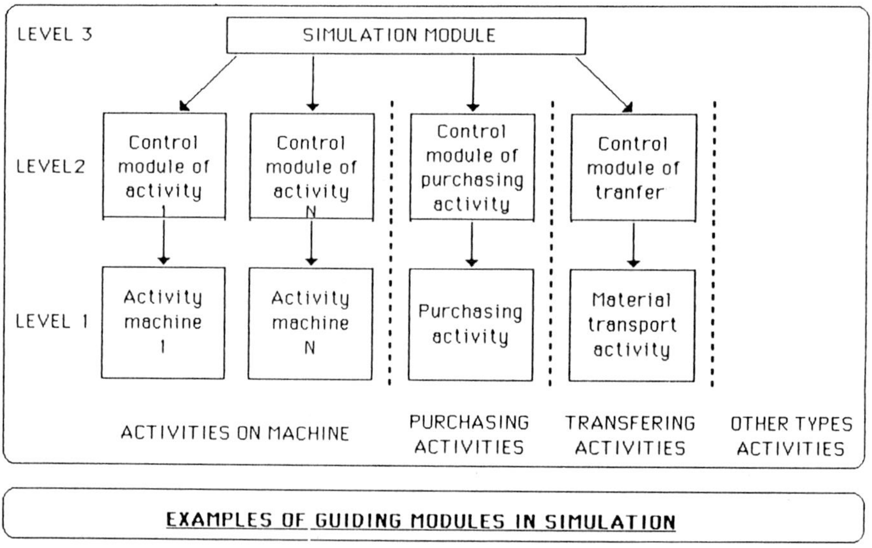

Figure 11

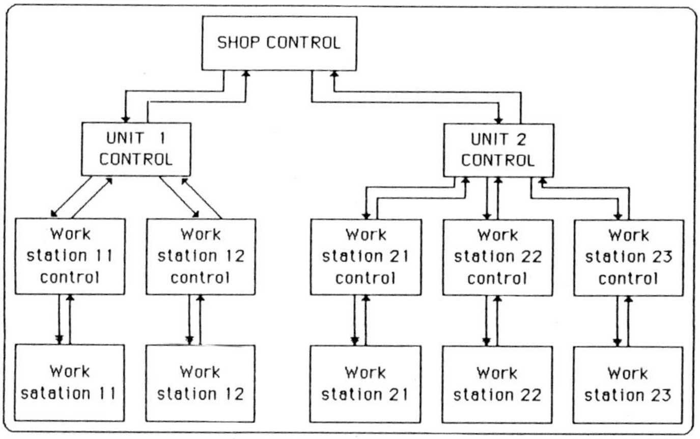

EXAMPLE OF A HIERARCHIC STRUCTURE OF MANUFACTURING SYSTEM WITH GUIDING MODULES

Figure 12

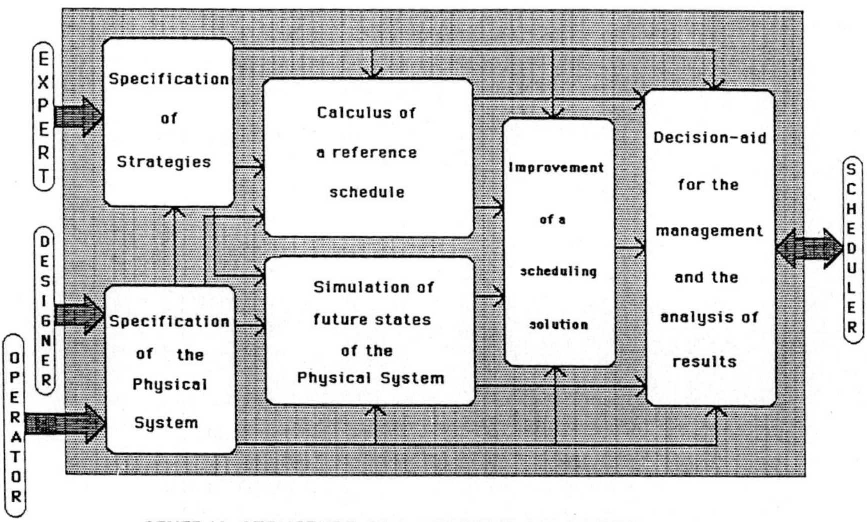

GENERAL STRUCTURE OF A DECISION-AID SYSTEM FOR SCHEDULING AND GUIDING A WORKSHOP

Figure 13

REFERENCES :

[DOU 84.11] G. DOUMEINGTS - "Méthode GRAI : Méthode de conception des systèmes en Productique" - Thèse d'Etat, Université de Bordeaux 1 (November 1984).

[DOU 85.3] G. DOUMEINGTS - "How to decentralize decisions through GRAI model in Production Management" - IFIP W.G. 5.7 International Working Conference, Munich, RFA (March 1985).

[SIM 60] H.A. SIMON - "The new science of Management Decision" - Harper and Row Publisher, New York and Evanston (1960).

[SIM 69] H.A. SIMON - "The science of the Artificial" - MIT Press (1969).

[SIM 75] H.A. SIMON - "Administrative behaviour. A study of decision making process in administrative organization". Third Edition (1945, 1957, 1975). The free Press, New York (1975).

[MES 70] M.D. MESAROVIC, D. MACKO, TAKAHARAY - "Theory of hierarchical multi-level systems" - Academic Press, New York (1970).

[MEL 72] J. MELESE - "L'analyse modulaire des systèmes de gestion". Editions Hommes et Techniques (1972).

[PUN 84] L. PUN - Systèmes industriels d'intelligence artificielle - outils de Productique" - Editions : Editest in french (1984), Plenium in english (1986), Friendship in chinese (1986).

[DOU 84.8] G. DOUMEINGTS, C. BERARD, D. BREUIL, MC MAISONNEUVE - "Use of Artificial Intelligence for piloting manufacturing unit" - International Working Conference on "Modelling of Production Management Systems, PMS" - Copenhague (DK) (August 1984).

[BER 85.8] C. BERARD, V. BRAUD, G. DOUMEINGTS - "Guiding techniques for manufacturing systems using Artificial Intelligence" - ICPR, Stutgart, RFA (20-22 August 1985).

PROCESS CONTROL

ARTIFICIAL INTELLIGENCE IN MANUFACTURING
T. Bernold (editor)
Elsevier Science Publishers B.V. (North-Holland)

PILOTEX: EXPERT SYSTEM FOR PROCESS CONTROL AND MAINTENANCE

Jean-Michel HUSSON
I.T.M.I.
Chemin des prés
ZIRST 38240 MEYLAN
FRANCE

ABSTRACT

PILOTEX is a generic expert system (ES) for industrial process control and maintenance. It is used to build applications consisting of:

- Standard modules (inference engines, a data base...),
- Specific modules (human and computer interfaces...),
- Knowledge bases.

PILOTEX is designed to help manage complex systems for which it is impossible or too costly to define a valid mathematical model. The different inertia present in these systems mean that time and external connections are primordial factors for all associated operations.

PILOTEX is developed by I.T.M.I. and has already been implemented in industrial (control, maintenance) and military (command assistance) contexts. Therefore, a set of human and software interfaces is available to improve testing and integration stages of the ES.

KEYWORDS: Industrial expert systems, Temporal knowledge representation, Process control and maintenance.

1 - WHAT IS PILOTEX?

PILOTEX is a generic expert system for industrial process control and maintenance designed and commercialized by ITMI, a French company based in Grenoble. From the very beginning of its development, PILOTEX has been designed for industrial applications and improved with the feedback of users and experts. PILOTEX deals with the industrial world, therefore numerous tools have been designed concerning the various aspects of development and maintenance of the knowledge used by the system.

2 - WHERE CAN PILOTEX BE USED?

PILOTEX helps to manage complex systems for which it is impossible or very costly to define a valid mathematical model. This applies in particular to production plants in the fields of metallurgy, chemistry, paper or glass production, oil refining... or in the military field for command assistance. Frequently, industrial systems are based on a closed-loop model:

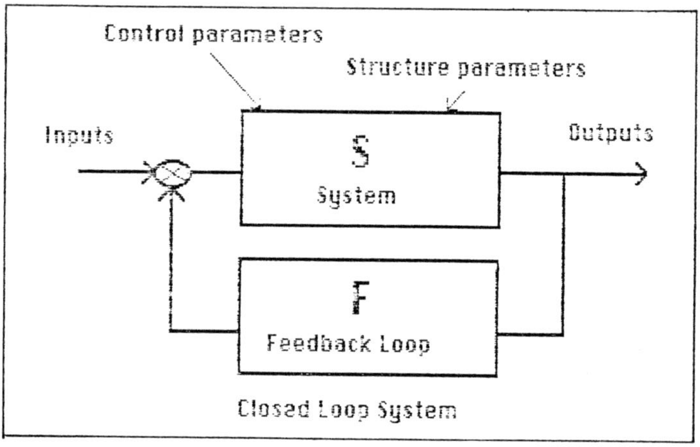

The formal theory of closed-loop systems presents numerous limitations when dealing with the real world:

- The system should be linear,
- Its limits have to be well-defined,
- A precise mathematical model should be established.

For complex or large systems, it is usually very difficult or impossible to reach such a mathematical description, but, despite this, experts use concepts and methods developed for this theory:

- Determination of a state for the system
- Distinction between control and structure parameters
- Methods of measurements
- Importance of "temporal concepts" such as inertia, evolution, delay, transition, stabilization...
- Action planning to reach an objective "smoothly",this objective being expressed in terms of state.

The mathematical models of the system theory are thus replaced by empirical knowledge and, occasionally, by algorithms solving partial aspects of the overall problem. This knowledge is acquired by the expert after a long tutoring where the "learning expert" is teamed with a skilled one and builds, little by little, his OWN methods of monitoring the system. Therefore, a usual situation consists in a group of experts, sometimes dedicated to supervising certain aspects of the system (e.g. starting, ending, day to day management, exceptional cases), and of whom each has his own way of addressing the problems.

All these facts are taken into account by the technology of expert systems and solutions can be found to teach, manage, unify, develop and formalize the knowledge of the experts.

3 - WHY AN EXPERT-SYSTEM?

3.1 Management

Industrial process management is the first area where the use of an expert-system (ES) such as PILOTEX fully applies:

- The ES can *handle quickly a large amount of data* and thus avoids forgetting information, especially when a decision is to be taken quickly. Helped by the ES, the expert can focus on the high level reasoning, leaving data collection, analysis and first reasoning to the ES.

- The ES is *available 24 hours a day*, 7 days a week thus reducing the well-known delays during which people wait for the expert's advice or avoiding the famous "Friday decision" taken to get a quiet week-end. The consequence of such behaviors is a lower productivity for the system since it is not tuned as efficiently as possible .

- *Homogeneity* of monitoring is a real problem in continuous processes: the first action of a foreman when he starts his shift is to change the parameters tuned by his colleague 8 hours before. This practice forbids medium term actions where a plan is to be followed over several shifts. PILOTEX, being able to produce AND EXPLAIN such plans, is very profitable since large drifts are avoided and the explanation is of primary importance if the human supervisor is to accept the proposed plan.

3.2 Diagnosis

The ES is also helpful for diagnosis and alarm procedures but several aspects of this problems coexist:

1) The system is directly driven by the ES when the problem is to take account of a *large number of alarms during a "short time"*. This type of situation is called "real-time" and lead to special programming techniques to allow the program to be ready when new events occur in the process. The trouble with ES is the difficulty of taking into account new events in the middle of a deduction. Actually, commercialized ES, and this applies to PILOTEX also, do not really solve the problem: the only solution is to get a sufficient speed through hardware and software techniques to allow deduction to be finished before a new event is allowed to occur. Some priority considerations can be made to delay the analysis of less important alarms if higher priority alarms occur.

2) The alarms are not time-critical and the main goal of the ES is to *propose a diagnosis*, and sometimes explanations and solutions, to an abnormal behavior of the process. For this type of problem, if the ES has a sufficient internal time representation, it can elaborate difficult diagnoses (the feedback loop brings many of the difficulties) and eventually provide solutions (in closed-loop systems, knowing the cause of an abnormal behavior does not give automatically the solution to this behavior, sometimes the cause can disappear by itself without solving the problem). PILOTEX can, for instance, detect not only an *abnormal evolution* of the process which needs an adjustment of the control parameters but also *divergences* where the reality does not match with the model the ES uses at a certain time. For instance, if one heats a room in which windows are opened, a normal regulation scheme would never work and,worse, would lead to an important energy waste.

3.3 Teaching

The education of new experts can be improved in quality and time with an ES: by storing *typical and exceptional scenarios*, the learning expert will encounter more quickly a wider set of cases for which all records are available and for which interactive explanations can be given to explain the decision making process. The ES must propose (and that is true for the 2 former points) not only an accurate description of the expert knowledge but also the possibility to get commentaries and messages about this knowledge and its use in a line of reasoning.

3.4 Research

An ES can be used to test new ways of managing the system without risking a trial-and-error behavior in reality. In this kind of application, the ES provides tools allowing *automatic testing of new knowledge* with a set of typical or exceptional scenarios, the ES is also able to provide statistics and debug information about new items of knowledge.

4 - HOW DOES PILOTEX WORK?

4.1 A rich expert language

A key fact about PILOTEX is that it proposes a large set of possibilities to describe the expert knowledge. The first task for the expert is to give information about the *"working universe"*. This is achieved by using the following concepts:

- *Element*: an element is a basic fact for PILOTEX, it has a name, a value of a certain type (boolean, integer, real, array, scalar...), an *"access strategy"* to determine the value, a *"production strategy"* to eventually act on the external world, a temporal description consisting of a sampling period and a lifetime which will allow the handling of both synchronous and event-based time.

- *Set:* a set is a static or dynamic collection of elements. For instance, the set of measurements can statically contain the elements temperature, speed, acceleration... and the set of alarms can contain all the measurements out of the standard limits, this being dynamically determined during the "reasoning phase".

- *Functions*: in the fields of industrial processes, a lot of algorithms have been developed representing items of knowledge, why should an expert rewrite them ? PILOTEX gives the possibility of writing "internal" and "external" functions. Internal functions are written in LISP (a large library is already available with PILOTEX). External functions can be written in any language as soon as the parameters passing between PILOTEX and such an algorithm is written in accordance with the (simple) standard developed by ITMI. These external functions are generated by PILOTEX as independant tasks resulting in powerful techniques of information exchange between PILOTEX and the external world.

Rules: Finally the expert uses rules to express his knowledge. These rules consist in an "IF-THEN-ELSE" part:

IF <condition> THEN <conclusion> ELSE <conclusion>

The condition being a combination of expressions which can contain *quantifiers*, operations on elements and sets, and functions calls.

A rule can also contain an optional *"application set"*, e.g.

FOR X IN ALARM_SET (IF DURATION(X) > 3 days THEN ...)

giving to the expert a very neat and powerful way to write the equivalent of several rules in a classical ES.

Finally, the rules are gathered in *rule bases* corresponding to the main steps of reasoning (see 4.2 below).

4.2 Multiple inference engines

A special module of PILOTEX is dedicated to the sequencing of basic steps of reasoning, for instance, a very usual sequence of reasoning for closed-loop systems is:

- State evaluation
- Comparison between current evaluation and previsions
- IF state evaluation does not agree with previsions
  THEN - Find why
  - Establish a new plan or correct the current one
  ELSE continue the current plan
- Prevision updating

In real implementation, this sequence contains usually 12 to 15 steps containing measurement checking, alarms response at any step etc... For each step, PILOTEX activate the corresponding rules base with the necessary strategy (data or goal-driven strategy).

4.3 Tools for knowledge base debugging and knowledge management

Debug information (trace, breakpoint) and statistic information (set of "programmable counters") can be used by the expert for any item in the knowledge description. PILOTEX can *repeatedly perform expertises* on a set of rules to be tested ; can be used in conjunction with *test files* containing typical or exceptionnal scenarios ; can work in non-interactive sessions allowing the *chaining of expertises* especially useful for knowledge base validation . Moreover, the expert can *redirect* input-ouputs, error and trace messages on any device available on the computer.

These tools are of first importance for the user to reduce costs linked to writing, testing and managing knowledge.

4.4 External world connections

The expert can make PILOTEX *communicate with the external world* in both input and output, by means of external functions. Actually, PILOTEX uses only the *"polling technique"* to get information from the external world, the "interrupt technique" is at the stage of R&D at ITMI and will be used to develop a *"REAL TIME PILOTEX"* where time-critical problems, as presented in 3.2.1, will be solved in a satisfactory manner.

4.5 Powerful interfaces

Human interfaces will be a very important point in the success or failure of an ES. Innovating from the current trends, ITMI develop *two kinds* of interfaces:

- The *user* interface, as simple as possible, developed on graphics displays with the well-known mouse, menus and windows tools and the intensive use of drawings. *The user is not be supposed to be an expert* and does not have the same goal as the expert when using the ES. This fact must be taken into account when developping an ES for the industrial world.

- The *expert* interface is *very difficult to implement*. The writing of knowledge should be as easy as possible and that mean a trade-off to be found between the completely directed editor often too cumbersome to use and the completely free editor which forces the expert to learn a complete language. The manipulation of the available tools for debugging and testing should also be as easy as possible but the first step in that direction is to determine and build these tools. From lexical checking to dynamic inconsistency resolution, many functions can be implemented which are not of equal importance. For these reasons, and because of its industrial experience, ITMI focuses, at this point in time, on the *quality of expert tools for debugging and maintenance* rather than proposing to the expert the possibility of opening tens of windows on a gigantic display, each of which contains a small text or diagram.

5 - WHEN WILL PILOTEX BE AVAILABLE ?

PILOTEX has already been installed "off-line" for industrial applications. The version for "on-line" integration will be delivered in 86 and will be one of the very first expert-system, produced by an AI company, installed in the real world. This version will take account of the numerous remarks and adjustments made by experts during their use, over several months, of the "off-line version". Any information about PILOTEX can be obtained from:

ITMI
Chemin des prés
ZIRST 38240 MEYLAN
FRANCE
Tel: 76 90 33 81
Telex: 320003F

ARTIFICIAL INTELLIGENCE IN MANUFACTURING
T. Bernold (editor)
Elsevier Science Publishers B.V. (North-Holland)

PMS-1 A REAL-TIME ADVISOR FOR THE BORON CONTROL IN PWR NUCLEAR REACTORS

J. GUILLERMARD and J.C. LAGACHE

FRAMENTEC S.A.
La Boursidière - R.N. 186
92357 LE PLESSIS ROBINSON CEDEX
FRANCE

The goal of the PMS (Plant Monitoring System) project is to apply A.I. techniques in building a real-time advisor for the control of a PWR nuclear reactor. The PMS-1 prototype is focused on the control of the boron concentration in the primary circuit of a PWR.

PMS-1 is built around a blackboard model using both the event driven and the goal directed paradigm. The knowledge base, that has been developed by connecting the system to a PWR simulator, includes the numerical model of an automated device and a set of heuristic rules. The system analyzes data sent every 30 seconds through a RS232 by the simulator.

This paper discusses the approach followed when developing the system, including the description of a test case that exhibits the different actions recommended by PMS-1. Finally, it suggests future steps toward an actual "intelligent" monitoring system.

1. INTRODUCTION

The nuclear industry provides many potential applications for the expert system technology. Among all of these, the category of real-time problems is probably the most challenging. In this area, one of the ultimate goals is to build multi-purpose "intelligent" monitoring systems, both for normal and accidental operation of nuclear power plants. As a first step toward this goal, Framatome, the French nuclear company and its knowledge engineering subsidiary Framentec started the PMS-1 project.

The main goal of the PMS-1 project is to demonstrate the feasibility of applying A.I. techniques for building a real-time advisor. The purpose of this paper is to describe our approach of the problems and the results obtained during the first phase of that project.

The way we proceeded can be summarized as follows:

* First a subset of the control of the PWR nuclear power plants was chosen, namely the control of the boron concentration in the primary circuit of the plant (see Appendix I). This subset was chosen because:

  - boron concentration regulation is an important operation (see Appendix 1)

  - the time constant for the boron monitoring is not too small (sensors acquisition every 30s is a reasonable time interval)

* Second, instead of developing PMS-1 on a nuclear plant, we decided to work on a nuclear simulator that reproduces the behavior of a nuclear plant in an environment similar to a real control room (see Appendix II for the description of the SAF simulator).

* Third, a knowledge engineering tool was chosen. Since it appeared necessary to combine both event-driven and goal based paradigms, we decided to use a modified version of AGE (1).

2. PMS-1 Architecture

From the knowledge engineering point of view, the basic requirements for the prototype were:

* the availability of both the event-driven and the goal-directed paradigms

* the feasibility of connecting the system to a numerical simulator to perform synchronous data acquisition

* the feasibility of running the knowledge engineering tool in a real-time environment

* the possibility of designing a window oriented user interface

2.1. Hardware Configuration

Considering the software choices (modified AGE + SAF simulator), the hardware configuration is based on the three following elements:

* a Gould 32/8780 to run the simulator

* a Xerox 1108 to run the knowledge engineering tool and the knowledge base

* a communication link between the Gould and the Xerox

At this stage of the project, very little efforts have been dedicated to the hardware problem. Since the goal is first to demonstrate the feasibility, we looked only for a simple and robust link and then chose, for the moment, a RS232 link.

Whereas sophisticated communication protocols (such as the Gould sends data only when it gets a "ready to transmit" signal coming from the Xerox or two parallel tasks run on the Xerox, one for the system the other waiting for input from the Gould) were tried, PMS-1 uses a simple one-direction transmission system.

Every 30 seconds, a process, that runs on the Gould 32/8780 in parallel with the simulation program and shares part of its memory, sends data to the Xerox 1108 through the RS232 link. On the Lisp machine, as soon as data are received, the knowledge system analyzes these data, makes recommendations and waits for new data. Since the analyzing cycle of the knowledge system always lasts less than 30 seconds, there is no simultaneity problem. Explanations are produced on the Xerox 1108 by a low priority process that will be activated only if the main process is not running out of time. This is done by comparing the time spent since the data were received with the progress made.

2.2. Knowledge Engineering Tool

The tool used is a modified version of AGE running on the Interlisp-D environment. Modifications to the AGE software were required for combination of the forward and backward chaining paradigms and also for memory management.

2.2.1. Design Options

The basic forward-chaining framework is used. The LIFO strategy for event selection is used. In addition to that, backchain knowledge sources are called from inside event-driven knowledge sources. As backchain knowledge sources are explicitly called from event-driven knowledge sources, there is no need to use the stack of goals and thus the focus of attention is reduced to the choice of the next event to proceed. So far, no certainty factors are included in PMS-1. The AGE explanation mechanism is not used by PMS. Instead, a specific graphics-oriented explanation mechanism has been designed.

2.2.2. Specific design for real-time requirements

The amount of memory used by the system has to remain constant during run-time. For this purpose, all instances (or nodes in AGE terminology) of hypothesis levels are created during the initialization phase.

In the same way, the event stack is cyclically refreshed because the system doesn't need to keep track of all events and only the latest events are used for the reasoning.

2.3. Knowledge Base

2.3.1. Structure of the Knowledge Base

The knowledge base includes:

* A set of knowledge sources used for initializing the system (creation of the geometry, creation of the history ...)

* A set of knowledge sources that are used in temporal cycles of reasoning. A reasoning cycle itself can be divided in three parts:

  - The numerical part that modelizes an automated device that controls the boron concentration

  - The heuristic part that criticizes the result given by the automatic device

  - The action part that decides and displays how to perform the actions advised by the heuristic part (depending on the circuit configurations). The explanation module is included in this part.

The simple (event-driven, LIFO) paradigm is used for all knowledge sources except for the heuristic part which is represented by a backchaining knowledge source. Both parts share the same attributes in the hypothesis structure.

2.3.2. The hypothesis Structure

The hypothesis structure can be divided into three parts :

* Modelization of the geometry of the power plant. Various attributes represent the properties of the components. Only the components of the plant that are relevant to the boron monitoring are represented in the PMS-1 knowledge base. In the hypothesis structure, the relevant levels are CORE, CVCS (chemical and volumetric circuit), PIPE (output and input legs), and PRESSURIZER. The topology of the circuit is represented by links that connect nodes representing the components.

```
CORE -
        MEASURED.POWER - (100)
        MEASURED.TOP.POWER - (45.5)
        MEASURED.BOT.POWER - (54.5)
        MEASURED.R.BANK.POSITION - (211)
        MEASURED.GREY.BANK.POSITION - (625)
        DEMAND.GREY.BANK.POSITION - (625)
        BORON.CONCENTRATION - (552)
        DELTA.PHI.REF.100 - (-6.15)
        CONNECTED.TO - (COLD.LEG.3 HOT.LEG.3
            COLD.LEG.2 HOT.LEG.2 COLD.LEG.1
            HOT.LEG.1)
```

The CORE level

* The set of levels that can be described as computation nodes (i.e. where results of symbolic and numerical computations are stored). The nodes SYMBOLIC.STATE, OPERATING.PARAMETERS, STATE. VARIABLES belong to that category.

```
SYMBOLIC.STATE -
    CURRENT.TIME - (29-May-85 13:58:47)
    PREVIOUS.TIME - (29-May-85 13:58:09)
    SYC.CURRENT.MODEL - (STAND.BY)
    SYC.PREVIOUS.MODEL - (STAND.BY)
    SYC.AREA.LOCATION - (DELAYED.BORATION.AREA)
    SYC.PREVIOUS.AREA.LOCATION -(DELAYED.BORATION.AREA)
    SYC.AREA.DURATION - (46)
    SYC.DILUTION.FLOW.RATE - (0)
    SYC.BORATION.REF.VOLUME - (0)
    SYC.BORATION.INJECTED.VOLUME - (0)
    SYC.BORATION.TIME.INTERVAL - (0)
    CURRENT.MOTION - (UNKNOWN)
    ACTION.DETERMINATION.PERFORMED - (29-May-85 13:58:44)
    LAST.ACTION - (BLOW.BORON.SLUG)
    LAST.ACTION.TIME - (29-May-85 13:57:23)
```

TIME.SINCE.LAST.ACTION - (46)
FORECAST.ACTION - NIL
CURRENT.ACTION - (STAND.BY)
PREVIOUS.ACTION - (STAND.BY)
ADVICE - (CURRENT ACTION IS STAND.BY)

The SYMBOLIC.STATE level

* The HISTORY level and the associated nodes (see paragraph on selective forgetting).

We have to point out that the nodes of the structure are not hierarchically organized in PMS-1. Semantically, one has to consider subsets of the list of all hypothesis levels instead of a stack of levels.

2.3.3. The Numerical Part of the Knowledge Base

For boron monitoring, a numerical automated device was available, but it was not optimal especially regarding the volume of effluents produced.

The automated device represented in the knowledge base includes filter, derivative and integrator modules. This part of the system is algorithmic and can be summarized in four steps:

* Data acquisition: the system acquires data from the simulator

* State variables computation : from the data acquired and the history stored, the system computes meaningful state variables (this includes the numerical computation of Laplace variable filters)

* Action determination: the system uses simple criteria on a diagram whose coordinates are the axial offset and the position of R control rods, to determine the action to perform (boration, dilution), if any

* Setpoint computation; if an action has to be taken, the system computes the associated setpoints (volumetric flow-rate, duration of injection ...).

2.3.4. The heuristic part of the knowledge base

The heuristic part of the knowledge base is encoded in the form of a backchain knowledge source. It contains rules that try to improve the conclusions made by the numerical part.

The possible improvements can be divided as follows:

* Anticipation of the action: considering trends and future transients, it may be interesting to perform an action before the automated device decides it. A good example is the following rule that will advice to start the dilution process even if the calculated recommendation is to wait:

```
(7 ((NOT.PMS.EQUAL SYMBOLIC.STATE PREVIOUS.ACTION CONTINUE.
    PRE.DIL)
    (PMS.MEMBER SYMBOLIC.STATE SYC.AREA.LOCATION
          (DELAYED.BORATION.AREA NEUTRAL.AREA))
    (NOT.PMS.PROFILE.PARSE OPERATING. PARAMETERS LOAD.
    PROFILE.33)
    (PMS.LESS STATE.VARIABLES LOCAL.VS.PAST.POWER -20)
    (PMS.EQUAL SYMBOLIC.STATE CURRENT.MOTION TOP.RIGHT.
    MOTION))
    (NIL evtype NIL
          element SYMBOLIC.STATE
          atval ((FORECAST.ACTION (LIST (QUOTE INITIALIZE.
          PRE.DILUTION))
                NIL ))
               chtype BCSUPERSEDE
               comment (%( %))))
```

Example of a heuristic rule

This rule basically states that even if the calculated advice is to take no action (SYC.AREA.LOCATION = DELAYED.BORATION.AREA or NEUTRAL.AREA), if there is no power increase forecast for the next 20 mn (LOAD.PROFILE), the power level is less than the average power for the last three hours minus 20 % (LOCAL.VS.PAST.POWER) and the trend on the motion of the RxDI point is top right motion (CURRENT.MOTION) then the dilution can be initiated.

* Delay of the action: in order to minimize the volume of effluents it may be advisable to wait before accomplishing an action even if the automated device recommends it

* Consideration of plant characteristics: the rules can also take into account factors, such as a bad calibration of the control rods, that are not considered by the numerical part.

When no improvement can be made, the systems uses as default the recommendations of the automated device.

2.3.5. The action part of the knowledge base

In order to take an action, one has to consider the current state of the circuits (which is directly dependent on the previous actions performed). Once it has been decided to accomplish a given action, the system will help the user to perform that action by:

* visualizing on a drawing of the circuits the valves that need to be used

* showing on a display of the panel of the simulator the switches that needs to be turned on or off.

This is done by storing the different ways to accomplish an action, depending on the current state of the plant described from the history of the actions performed previously.

Explanations are based on the firing of the rules contained in the backchaining knowledge source.

2.3.6. Selective Forgetting

Part of the hypothesis structure is dedicated to the representation of the history of the power plant operations. The basic assumption for the control of the boron concentration is that for every state of the power plant, the reasoning can be done with a limited history (i.e. a few hours) of the main operating parameters (power, axial offset ...).

A simple way of representing this history is to have an history node in the hypothesis structure. When initializing the system, the user must give the length of the history (i.e. the number of history instances to be created). A typical history length is 360 (associated with a time-step of 30 seconds, this represents a 3 hours history). During the real-time run, the system writes on history instances in a cyclical way.

```
HISTORY3 -
        PROTO - (HISTORY)
        TIME - (29-May-85 13:56:29)
        NIL - NIL
        POWER - (100)
        POWER.DERIVATIVE - (0.0)
        POSITION.R - (205)
        DELTA.PHI.REL - (-3.85)
        BORON.CONCENTRATION - (550)
        AREA.LOCATION - (BORATION.AREA)
        AREA.DURATION - (NIL)
        ACTION - (WAITING.BEFORE.NEXT.BORATION)
        INST - NIL
```

Example of an HISTORY node

Associated with this history, the knowledge base includes trends predicates that allow rules to test the tendency of crucial parameters.

3. THE USER INTERFACE

A window interface has been designed. All windows are passive except the EXPLAIN window that can be activated by the mouse during run-time. The main windows are:

* The typescript window where the system prints a summary of the status of the plant at each time step

* An action window that shows actions to be taken and the associated setpoints

* An explanation window where explanations are prompted

* A circuit window that shows the boration/dilution circuit and visualizes the actions to be taken

* A SAF panel window that shows the location on the SAF simulator for the buttons that correspond to the actions to be taken

* In addition, several windows display the time variation of the key parameters (power, R rods location, Boron concentration ...).

4. A TEST CASE

4.1. Description

A classical test case for the control of the boron concentration is a decrease of power, from a steady-state at nominal power, to a steady-state at 50 % of nominal power. A typical slope for such a power transient is 15MW/mn (which means about half an hour for a 900MW reactor when decreasing the power level from 100% to 50%).

An example of the chronology of such a transient is the following:

* 15:46:05 start of power decrease

* 16:12:36 PMS-1 advice : anticipated dilution (flow rate 27 m3/h)

* 16:13:36 PMS-1 advice : dilution (27 m3/h)

* 16:18:36 end of power decrease

* 16:43:07 PMS-1 advice : boration (volume 80 I, flow-rate 9 m3/h)

* 16:54:07 PMS-1 advice : "flush" boron slug

4.2. PMS specific advantage on this transient

Compared with the numerical automated device, the specific advantage of PMS-1 is to be able to anticipate some actions or similarly to avoid taking unnecessary actions. In this particular transient, PMS-1 was able to anticipate the dilution. Later, it advised to "flush" the boron slug. The consequence is a minimization of the volume of effluents generated.

It is difficult to give a quantitative evaluation of the decrease of the volume ofeffluents generated due to the use of PMS-1. From a qualitative point of view, the advice of PMS-1 improves the smoothness of the transient.

5. CONCLUSION

The primary goal of the PMS-1 project, i.e. applying Expert Systems techniques in a given real-time environment, has been successfully met. However, one has to point out the following limitations of this prototype:

* A fair amount of the knowledge is of procedural nature. At this stage, the heuristic knowledge is limited

* The interface between the simulator and the Lisp machine, although sufficient for the prototype, would not allow the transfer of large amount of information

* This methodology cannot be applied in a domain where the time constants are too small.

Nevertheless, PMS-1 is a first stage toward a real time advisor, that was successfully connected to a time dependent process, followed and helped controlling that process.

The next step to be investigated is to build a monitoring system that performs the actions instead of just advising the user of the system. In addition to that, we think that this methodology developed for PMS-1 can be applied in the domain of the post-accidental studies for PWR nuclear reactors. On a more general basis, it clearly appears that this kind of approach can also be used profitably for a large number of industrial process control problems.

REFERENCES

(1) AGE reference manual, AGE-1 October 1981 Heuristic Programming Project, Stanford University N. Aiello, C. Bock, H.P. Nii, W.C. White

(2) VM : representing time-dependent relations in a medical setting Stanford University, June 1980 L.M. Fagan

I. PWRs" Control Systems

Nuclear reactor power control systems have two major operational functions: they are generally described as regulation and shutdown. The regulation function involves mainly the short-term adjustment of the reactor power to the desired operating level and maintenance at that level despite transient changes in reactivity. These changes may be caused by internal or external factors such as reactor water temperatures variations or changes in steam demand. The control system must also have the capability to increase or to decrease the reactor power level, either gradually during normal operation or very rapidly in an emergency situation.

In Pressurized-Water Reactors (PWRs), two complementary methods are used for controlling the reactor: control rods and shim control. Control rods, containing neutron absorber materials, are combined in clusters of 20 or 24 rods which are inserted (or withdrawn) from the top of the reactor vessel. Fifty or more rod clusters are distributed throughout the reactor core. They are arranged in symmetrical groups (or banks) and all the clusters in each bank are moved simultaneously. Shim control is achieved by the chemical control system. Boron, in the form of boric acid is dissolved in the reactor water. As the reactor operates and the fuel reactivity decreases, the boron concentration is gradually reduced by passing the water through an ion-exchange system for removing boric acid (dilution).

A subsidiary function of the control system is to shape the neutron flux (i.e. to bring about local changes in the flux) in order to achieve the optimum power distribution in the core. Insertion of control rods at appropriate locations will help minimizing flux distortions. But the uniform distribution of the boron poison at the proper concentration throughout the core can also minimize the need for regulating-rod insertion and the consequent spatial distorsion of the power.

II. Description of the SAF Simulator

The "SAF" nuclear power plant analysis simulator, in operation at Framatome since April 1984 enables complex accidental and normal scenarios to be analyzed in detail.

The SAF is a multipurpose engineering simulator. It includes a computerized control room and a computer center built around a Gould 32/8780 bi-processor computer and Sintra Triade 20 graphics processors.

In the SAF, all the systems of a nuclear power plant are simulated.

* Core model

  - neutronics

    axial model (18 zones)
    1 D diffusion code
    2 energy groups
    single-point kinetic model
    simulation of all burnup values up to cycle 4

  - thermal

    special efforts have been devoted to core-reflooding models

* Reactor coolant system model
  A thermal-hydraulic model with separated phases and thermodynamic unbalance (24 nodes) is used. It includes:

  - a drift flux model in each link

  - interface exchanges in each node

  - dependence of reactor coolant pump head and friction coefficients upon two-phase flow

  - use of all models and correlations (including critical heat flux computation) normally included for design programs

* Steam generator model
  Steam generators are represented by a two point model with phase separation, permitting simulation of stratified SG operation with little or no steam production. It is qualified for SG tube ruptures, steam line breaks and feedwater line breaks

* Containment model
  The model includes separated phases and thermodynamic unbalance (superheated steam possible).

LIST OF SPEAKERS

Biedermann Jörgen
Dr., Comor AG, Computer- und Organisationsberatung,
Seefeldstrasse 62, CH - 8008 Zürich

Bitran Gabriel
Prof. Dr., Sloan School of Management, MIT,
15 Memorial Drive, Cambridge, Mass. 02139 / USA

Burkhardt Friedrich
Dipl.-Ing. FH, Ex-Cell-O-Corporation GmbH.,
Salacherstrasse 93, D - 7332 Eislingen/Fils

Cornwell Peter
Manager of the UK Manufacturing Systems Research
Group, Texas Instruments Ltd., M/S 46A, Manton
Lane, GB - Bedford MK41 7PA

Doumeingts Guy
Prof. Dr., Laboratoire d'Automatique, Université
de Bordeaux I, 351 Cours de la Libération,
F - 33405 Talence Cedex

Farinacci Marha L.
Staff Computer Scientist, Process Control and
Computer Technology Division, Alcoa Laboratories,
Aluminum Company of America, Alcoa Center,
Pittsburgh, PA 15069 - USA

Fox Mark S.
Prof., Director, Intelligent Systems Laboratory,
The Robotics Institute, Carnegie-Mellon University,
Pittsburgh, PA 15213 / USA

Franz Bernhard
Vertrieb Expertensysteme, Nixdorf Computer AG,
Postfach 40 12 20, D - 8000 München 40

Gairola Arun
Dr., Principal Engineer, Digital Equipment
International, Sudetenstrasse 5, D- 8950 Kaufbeuren

Gerelle Eric G.R.
CIM Marketing Manager, Digital Equipment Corporation
International (Europe), 12, Avenue des Morgines,
CH - 1213 Petit Lancy 1 / GE

Ghani Noordin
Technical Director, MARI, Microelectronics Ltd.,
32 Grainger Park Road, GB - Newcastle upon Tyne NE4 8RY

Guillermard Jacques
Knowledge Engineer, Framentec S.A., La Boursidière
R.N. 186, F - 92357 Le Plessis Robinson Cedex

Haton Jean-Paul
Prof., CRIN, Université de Nancy I et INRIA,
BP 239, F - 54506 Vandoeuvre

Husson Jean-Michel
ITMI-Industrie et Technologie de la Machine
Intellitence, Zirst, Chemin des Prés, F - 38240 Meylan

Jonckers Viviane
VUB, AI Laboratory, Plein Laan 2, B - 1050 Brussels

Kiratli Gisela
Dipl.-Inform., Laboratorium für Werkzeugmaschinen
und Betriebslehre, RWTH Aachen, Steinbachstrasse,
Blox 53/54, D - 5100 Aachen - Melaten Nord

Knasel Michael T.
Vice President and General Manager, Automated Systems
Division, MK-Ferguson Company, One Erieview Plaza,
Cleveland, Ohio 44114 / USA

MacConaill Patricia A.
Coordinator of Computer Integrated Manufacturing,
Commission of the European Communities, Information
Technologies and Telecommunications Task Force,
Rue de la Loi 200, B - 1049 Brussels

McArthur Robert C.
Arthur D. Little, Inc., Acorn Park,
Cambridge, Mass. 02140 / USA

Neumann Bernd
Prof. Dr., Universität Hamburg, Fachbereich
Informatik, Schlüterstrasse 70, D - 2000 Hamburg 13

Schneider Hans-Jochen
Prof. Dr., Technische Universität Berlin, Institut
für angewandte Informatik, Fachgebiet computerge-
stützte Informationssysteme, Franklinstrasse 28/29,
D - 1000 Berlin 10

Schraft Rolf D.
Dr.Ing., Fraunhofer-Institut für Produktionstechnik
und Automatisierung, Nobelstrasse 12, D - 7000 Stuttgart

Silverman Gary J.
International Business Machines Corporation,
11601 Wilshire Blvd., Los Angeles, CA 90025-1738 / USA

Smithers Tim
Dr., Edinburgh Project Manager of the ALVEY Demonstrator
Edinburgh University, Dept. of Artificial Intelligence,
Forest Hill, GB - Edinburgh/Scotland

Stotko Eberhard
Dipl.-Ing., Manager of CIM Programme Office IBM Europe,
Pocci Strasse 11, D - 8000 München 2

Timmermann Gerd
Leiter Gruppe Systemanalytik, Symbolics GmbH.,
Frankfurter Strasse 63-69, D - 6236 Eschborn/Ts.

Viti Paolo
Director Design Division, HQ Ing. C. Olivetti & C. spa.
Via Meravigli 12, I - 20123 Milano

Witkowski Mark
Texas Instruments Ltd., M/S 46A, Manton Lane,
GB - Bedford MK41 7PA

Yoshikawa Hiroyuki
Prof. Dr., Dept. of Precision Machinery Engineering,
Faculty of Engineering, University of Tokyo,
7-3-1 Hongo, Bunkyo-ku, Tokyo / Japan